NORTH AMERICAN WILDLIFE

NORTH AMERICAN WILDLIFE

David Jones

whitecap

The information in this book is true and complete to the best of our knowledge. All recommendations are made without guarantee on the part of the author or Whitecap Books Ltd. The author and publisher disclaim any liability in connection with the use of this information. For additional information please contact Whitecap Books Ltd., 351 Lynn Avenue, North Vancouver, BC V7J 2C4.

Edited by Elizabeth McLean
Scientific review by Daniel J. Catt, Dipl.T., B.Sc., M.Sc., RPBio.
Photo editing by Antonia Banyard and David Jones
Cover design by Antonia Banyard
Interior design by Tanya Lloyd
Front cover photograph by Thomas Kitchin/First Light
Back cover photograph by Peter McLeod/First Light

Printed in China by WKT

National Library of Canada Cataloguing in Publication Data
Jones, David (David Richard), 1956–
   North American Wildlife
   Includes index.
   Hardcover—ISBN 1-55110-900-X
         ISBN 978-1-55110-900-8
   Paperback—ISBN 1-55285-764-6
         ISBN 978-1-55285-764-9
   1. Zoology—North America. 2. Zoology—North America—Pictorial works.
I. Title
   QL151.J66 1999   591.97   C99-910846-8

The publisher acknowledges the support of the Canada Council for the Arts and the Cultural Services Branch of the Government of British Columbia for our publishing program. We acknowledge the financial support of the Government of Canada through the Book Industry Development Program for our publishing activities.

What is life? It is the flash of a firefly in the night.
It is the breath of a buffalo in the wintertime.
It is the little shadow which runs across the grass
and loses itself in the sunset.

Crowfoot's Last Words (1890)

Contents

Introduction

The fables of Aesop feature the most familiar animals of Europe and North America to impart their morals, but one of them, the grasshopper and the ant, is as much a parable for wildlife as it is for human beings.

An ant was stockpiling corn for the coming winter when a grasshopper happened by, gamboling and singing. The grasshopper stopped to ask the ant how he could waste such a fine summer's day in toil. The ant thanked the grasshopper for his concern, but replied that it was in his nature to be prepared. "Suit yourself," said the grasshopper, who hopped merrily off. In the traditional ending to the fable, the ant, snug in his winter larder, answers a faint knock at his door in the dead of winter to find the grasshopper on his doorstep, begging for any morsel the ant might spare. The ant rebukes the grasshopper for his lazy ways and turns him back out into the snow to face certain starvation.

A stand-up comedian popular twenty years ago invented his own ending for the tale, which, roughly paraphrased, was "Then winter came and they both froze to death." The comedian was the aptly named Jonathan Winters, and his version of the fable has a core of truth to it. In the tropics, food and water are abundant year round and life is limited only by other life. For every species there is another striving to eat it, infect it, or take the food from its mouth.

ABOVE: *A sunflower attracts flies and other insects, luring food to this tree frog. Rather than actively pursuing prey, most wait for targets to crawl or fly within range of their sticky, protrusile tongues, which unfold from the rims of their lower jaws.*

LEFT: *Most hoofed animals elude their predators with speed, but mountain goats have taken a different escape route, one so hazardous that few predators can follow. The goats pick their way along the high rock walls of mountains and canyons with the aid of narrow bodies, superb balance, keen eyesight, and rubbery hooves.*

Predation, disease, and competition also pose difficulties for temperate species, but north of the 35th parallel, all animals face a common imperative: After a summer of plenty, the hammer of winter falls.

The mammals have bartered a solution to the problem of winter with the currencies of milk and fat. With fat, they can draw on the summer's abundance over the lean winter. With milk, they pass this abundance on to their offspring at a time when newborns are too small or weak to forage for themselves. Nowhere is this strategy more dramatic than in the marine mammals. A northern right whale's body is almost 40 percent blubber. The mother hooded seal transfers blubber to her pup with the urgency of an aerial refuelling. Her milk is half fat, enabling the pup to double its birth weight before being weaned in only four days.

Many rodents forego the conversion to fat and store their food directly, stockpiling seeds, nuts, or grasses for the colder months. Their bodies in turn become food for carnivores that hunt them through the winter.

Reptiles and amphibians cannot offer their offspring milk, but their eggs allow their young to inherit nutrients during their early development. Lizards such as the Gila monster and the chuckwalla store fat in their tails to help them through the dry season. But in general, cold-blooded animals are less effective at storing fat than mammals. Instead, they excel in their ability to reduce spending to bare minimums. At the northern extremes of their ranges, snakes and frogs must hibernate half their lives to survive.

In the deserts of the American southwest, the availability of food is limited not by cold, but by water. Here, the reptiles and amphibians have excelled. Despite a complete dependence upon water during the larval stage of their life, some toads estivate through desert heat, cocooned in mud and mucus, waiting for the rains that will allow their eggs and tadpoles to swarm for a few weeks in temporary pools.

RIGHT: *The wild cats of North America are seldom seen. Cougars are now comparatively rare throughout most of the continent and lynxes live north of most cities and towns, but the bobcat, pictured here, is populous. There may be as many as 1.5 million of them living in the rural and wild areas of the United States.*

ABOVE: *Most North American mammals are asocial, living in close contact with others of their kind for extended periods of time only as juveniles. The black-tailed prairie dog is one of the few gregarious rodents. They live in groups of 7 to 12 animals known as coteries. Upon meeting, members greet each other with a hug.*

RIGHT: *The American alligator is vital to the ecology of the Everglades, not only as the top predator of the wetlands, but as creators of "gator holes." They make these thigh-deep wallows—3 to 6 metres (10 to 20 feet) across—by sweeping mud and vegetation from natural depressions in the bedrock. During the winter dry season, thousands of these holes serve as reservoirs for fish, amphibians, waterfowl, and all manner of aquatic life.*

Only in the southeastern corner of the United States do climatic conditions approach those in the tropics—stable, warm, and wet throughout the year. Not surprisingly, reptiles and amphibians are far more abundant here than in other habitats. The alligator is Florida's dominant predator, and the state has become a haven for hundreds of exotic species, many of them seeded by escaped pets.

The wildlife of North America are, in one sense, a remnant fauna. The mammals are the survivors of a wave of extinctions that occurred at the end of the last ice age, some 11 000 years ago. No one is certain what killed almost three-quarters of the large mammals at this time, but paleolithic hunters are suspected. One theory is that the large grazers, inexperienced with human predators, fell easy prey to these first colonists. The carnivores starved soon after. Reptiles and amphibians have survived a much earlier winter, one that wiped out the dinosaurs 100 million years ago, probably brought on by an asteroid or comet that struck the earth.

Like the ant in the fable of old, these survivors are the veterans in the battle for the calorie, which they must fight anew each season. It is a battle the animals can never win, but they can endure. This book is about their remarkable strategies.

ABOVE: *The wolf pack is a social hierarchy. There is some growling and fighting, but most of the behaviors that maintain this society are expressions of affection. Members of a pack greet each other by licking, nuzzling, and mouthing each other. Subordinate individuals may cringe and wag their tails. Many of the wolf's facial expressions are easily read by human beings—partly because many are innately similar to our own, and partly because of our long experience with domestic dogs.*

RIGHT: *This white-tailed buck is exhibiting a behavior known as lip-curling, or flehmen, from the German word meaning to implore or beg. The curling of the lip exposes the Jacobson's organ, a groove in the deer's palate that "tastes" airborne scents. Specifically, he is smelling for hormones in a nearby doe's urine that will tell him if she is entering estrus.*

LEFT: *Harbor seals are the most numerous of pinnipeds (seals, walruses, and sea lions) and range almost continuously from Baja north, through the Arctic, and south again to Cape Cod. Like most pinnipeds, mature harbor seal cows whelp only a single pup each year. But over the course of their long lives, which may be up to 40 years, they may produce many offspring.*

Committed Killers

A cat's whiskers just exceed the span of the rest of its body. With them, it can measure an opening into a rocky den in total darkness. At the touch of a leaf, it stops before making any rustling that might alert its prey. A cat's whiskers act as curb feelers. At least, that's the theory. Whether it's true or not, it seems a fitting adaptation for the cougar, bobcat, and lynx—wild cats prowling a narrowing evolutionary alley.

It is a path from which there is no turning for the world's most committed mammalian predators. They are carnivores, adapted for the procurement of meat and no other food. The proof of this specialization is in their skeletons, and in particular the sharp ends that poke through the gums. The canine teeth are the cat's killing tools. They are rich with nerves, and one theory holds that the cat is able to feel for the precise place to deliver its fatal bite.

Its teeth are at the end of a short lever—the cat's blunt muzzle—and are driven by powerful jaw muscles. In effect, cat jaws are like bolt cutters: Once they're positioned, few natural materials can resist this pressure. The saber-toothed cats, only recently extinct, were the greatest specialists of all. Their daggerlike canines were made for penetrating the thick hides

ABOVE: *The bobcat survives by ambushing a variety of small mammals, birds, amphibians, and reptiles. This one has caught a pheasant which, like most of the bobcat's prey, is smaller than the cat. But in parts of their range they also kill deer, animals that may outweigh the bobcat by seven times.*

LEFT: *Cougars may be the most efficient mammalian predators, killing as much as 80 percent of the prey they stalk. They are aided by acute vision, smell, and hearing, but the main reason for this high success rate is the manner in which they stalk. Cougars move almost invisibly, approaching their prey unseen to within a single bound before pouncing.*

ABOVE: *In this shot you can see the bowing action of the lynx's spine which stretches the length of its stride. While all of North America's cats are capable of bursts of speed, none can run for more than a minute. Prey able to elude a cat's initial pounce have a good chance of escaping.*

RIGHT: *Bobcats leave narrow tracks that sometimes appear to have been made by a two-legged animal. This may be the result of the bobcat carefully choosing its footfalls to travel silently when hunting. By stepping in the tracks of its forepaws with its rear feet, it minimizes the chance of betraying its presence with a sound.*

and bones of ground sloths, mammoths, and rhinoceros-like grazers. When these animals became extinct, the giant sabertooths that fed on them also died out.

At the sides of the cat's mouth are the carnassial teeth with which they separate flesh from bone. All of the carnivores—cats, dogs, bears, raccoons, and weasels—have carnassials, but the cats' are the sharpest. In all, their short muzzles accommodate only 28 to 30 teeth, far fewer than in other mammals of similar size. They have teeth for stabbing and slicing, and that's all.

Looked at crudely, the rest of the cat's body is a delivery system for those lethal canines. Cougars may lie in wait for their prey or prowl in search of it, covering up to 30 kilometres (20 miles) in a day, although 8 kilometres (5 miles) is more common. Other large predators may travel farther on their hunt, but the cougar does it while remaining nearly invisible. Even in the deserts of the southwest where vegetation is sparse, it moves using undulations in the topography as cover. In 1992, a cougar was tranquilized by wildlife control officers in the underground parking garage of the Empress Hotel in downtown Victoria, British Columbia. Victoria was then a city of 300 000 people, and nobody saw it

penetrate the city's core. The cougar's ability to move unseen is eerie.

A keen sense of smell, second only to that of the dogs and bears, alerts the cougar to the presence of a doe. Its large, swiveling ears help it zero in further in the dead quiet of snow-filled woods or among the insectile din of the Everglades, where perhaps 50 cougars—known as Florida panthers—still live. Like most predators, the cougar steals as close as possible to its prey before lunging, usually slinking in two or three short bursts. In the pauses between, it evaluates the animal with excellent eyesight. Although it does not see in color, its eyes are especially good at detecting movement, particularly in low light. A cat can see clearly in only 20 percent of the light that a person needs to discern objects. Cats usually hunt at night, when they have a visual advantage over their prey.

Cougars are among the world's top mammalian predators by two standards. First, they have a high kill rate compared to other cats. A cougar may successfully close 80 percent of all stalks with a kill. A lion, by comparison, feasts on only one in ten of the animals it stalks, a lynx maybe three out of ten. Second, cougars tackle larger animals than any other large solitary predator. A cougar taking down an elk may be subduing an animal eight times its size. They have to be more efficient than other cats simply because where cougars live, mammals of suitable size are much scarcer than they are on the African savannah.

Even so, the cougar, having crept close to its prey, must decide whether to attack or not. In the case of a deer, the decision would seem to be an easy one: The deer's teeth are suited to browsing. It has no claws, and its instinct is to flee rather than fight. But the deer is far from defenseless. The antlers of bucks can inflict wounds. And the hooves of a struggling deer can break skin or fracture a jaw.

Avoiding injury is important for predators, but especially for solitary

LEFT: *Cougar mothers care for their young for about two years. During their first few weeks she leaves them to hunt, bringing her kills back to their rocky den. Cougars have been known to bring their kittens live prey as small as grasshoppers on which to practice their hunting skills. Later, she leads the kittens to the site of the kill, and eventually takes them hunting with her.*

RIGHT: *At four to five weeks old, this kitten will investigate any small, moving object—a trait that remains with lynxes into adulthood. Trappers often lure them to a snare placed beneath a bit of ribbon or flagging tape tied to a branch. The fluttering of the fabric in the breeze is enough to attract the cat.*

OVERLEAF: *Cougars were once common throughout North America, and although their range is much reduced, they survive in an impressive range of habitats from the deserts of the American southwest to the coastal rainforests of the Alaska Panhandle. As the supreme carnivore, their killing tools remain effective wherever there is game large enough to stalk and kill.*

predators such as cats. Wolf skeletons have been found with fractures to the skull and jaws completely healed, indicating they can survive severe injuries. But for a cat, even if being kicked or gored by its quarry does not prove immediately fatal, the result may be the same without a pack from which to mooch during its recovery.

Having made its decision, the cougar attacks when it is either near enough to close the distance in a single lunge, or discovery forces it to bound after it. A pounce covers the gap between predator and prey silently, concluding with a stunning blow driven by the cat's body weight. A cougar's final lunge may cover up to 9 metres (30 feet) in a second.

The disadvantage of the pounce is that it's hit or miss. Once the cat is airborne, it has little control over where it will land, and if a deer hears the cat's takeoff, it has an instant in which to bolt—at right angles to the cat's direction, if it's lucky.

But a cougar going the wrong way isn't done yet. One reason for the cats' renowned agility is the linkages between their fine bones. The individual vertebrae of its spine are unconstrained by ligaments and the clavicle is vestigial. Instead, the bones are floating in muscle, which comprises a high percentage of the cat's body weight. Muscles acting on light, loose levers of bone make for extreme flexibility—the kind that allows a falling cat to twist to land on its feet. This same flexibility can shift a cat's direction the instant it lands.

A cougar that has missed its first lunge may choose to pursue a fleeing doe, but with each second that it doesn't close the gap, the chances of a successful kill diminish. Deer can run for five or six minutes without pause. But for a cat, one minute of running is a marathon. Afterwards, it may take 20 minutes to catch its breath. If the cat does manage to close the gap, the outcome of the hunt is still not certain. The cougar's five, needle-sharp claws emerge from the pads of its forepaws so that it can hold the neck or head of its struggling prey between them or pin the animal to the ground. Wolves rely on shock and blood loss to kill their prey, sometimes eating it alive, but cougars deliver a precise and lethal bite.

Because so few kills have actually been witnessed, there is some debate as to how cougars finish their prey. Some naturalists maintain that those sensitive canine teeth allow the cougar to pry apart two vertebrae and sever the spinal cord. Others argue that the cat simply suffocates the deer by crushing its windpipe.

RIGHT: *The cougar's only remaining eastern enclave is in Florida, where perhaps 50 Florida panthers, a subspecies of the cougar, survive. Almost all of these panthers have been caught and fitted with radio tracking collars, and despite intensive studies, they teeter on the brink of extinction. In contrast, there are nearly 700 permits currently issued in Florida allowing people to keep cougars as pets. Some biologists maintain that with the number of cougars that have escaped over the years, the purebred Florida panther must surely be extinct already.*

The deer, of course, is still doing its utmost to be one of the lucky two out of ten. One specialist in cat predation on deer and sheep estimates that one-quarter of all cougars, which live an average of eight to ten years, eventually die while attacking prey. Cougars riding on the backs of fleeing deer and elk have fallen off and broken their backs on tree trunks or followed their prey over cliffs. Others have been impaled by the spikes of branches and died of infection. If a buck can throw a cat, it has a chance to gore it with its antlers or fracture its skull with a hoof.

But in the usual course of events, the cougar is the victor. Having severed or crushed the spine, it may eat the animal on the spot or decide to drag it to a more secluded place. This is where the cat's astonishing strength is evident. A big male cougar weighs about the same as an average adult man, but is much stronger. They have been known to drag the bodies of cows three or four times their weight up to half a kilometre (a quarter mile) away. They can bound over logs with adult deer clamped in their jaws.

Once it has found a spot to its liking, the cougar gorges itself, sometimes putting away up to 8 kilograms (18 pounds) at a sitting. Often, it then covers the carcass with leaves or soil. It does so to reduce the odor and the chances of drawing bears or other scavengers. If game is scarce, the cougar may return the next day to feed again. There is one case of a cougar in Montana consuming an elk over a period of almost a month. Cougars have been known to sleep on their kills during cold weather, presumably to keep the carcass from freezing between meals.

All three of North America's wild cats are secretive animals. Naturalists and others who make their living in the bush may go their whole lives without ever seeing more than their tracks, but the lynx is especially shy, a creature of Canada's boreal forest and the deepest woods of the northern United States. It is a small, fine-boned cat. An adult female weighs barely 7 to 9 kilograms (15 to 20 pounds) and a male is about 40 percent larger. At this weight, it can stay afloat on its huge paws even in powder snow. Indeed, before the invention of radio collars and satellite transponders, much of what we know about lynxes and other cats came from the study of their tracks.

Nowhere is the feline commitment to meat-eating more evident than in the story of the lynx and the snowshoe hare—by now a proverb of ecology. Lynxes feed almost exclusively on snowshoe hares. When the hares prosper from an abundance of forage, the lynxes feed heavily and

begin having very large litters—as many as eight kittens. Their numbers grow rapidly, peaking one or two years after the hare population reaches a maximum. But as the hares decline, lynx numbers quickly follow. Since records of trappers' pelts were first taken in the 1820s, the time between peaks has been a consistent 10 years.

It was once thought that the hares ate all their food and then began to starve. The populations of lynx then began to plummet because they, too, were starving. Biologists now believe that hare numbers fall not only because of heavier lynx predation, but because plants begin producing toxins that inhibit the hares' ability to digest them, in response to heavy browsing. Lynx numbers fall because after hare numbers have peaked and they become harder to find, the lynxes start having smaller litters—sometimes a single kitten. Nonetheless, lynxes do suffer during a shortage of hares, and may turn to ducks, grouse, carrion, or even foxes if they can find them; others will even resort to cannibalism.

Adults and yearlings may migrate to areas where food is more abundant, but as the hare population cycle is synchronous throughout the north, seldom do these migrants find more hares. During the 1995 low, many lynxes emigrated from the southern Yukon to the Alaska Panhandle. Many others starved. The cycle persists today, although the peaks are much lower, a quarter of the numbers during the 19th century.

For the lynx, there is only one good reason to abandon its stealthy ways, and that is the opportunity to make more lynxes. The females pronounce their readiness to mate in late March or early April with a mournful wailing. The male her yowling attracts is usually from an adjacent territory, and copulation lasts only a few seconds—although it may be repeated many times in an hour.

Two months later, the female lynx gives birth to from one to eight

RIGHT: *Like all cats, bobcats have well-clawed paws for immobilizing their prey before delivering a killing bite. Depending on the size of its prey, the bobcat may sever the spinal cord or crush the animal's windpipe. Cats have fewer teeth than other carnivores, reducing the number of bite pressure points and increasing the force behind each.*

young, depending upon the snowshoe hare cycle, usually in a thicket or other secluded spot. They're born at the beginning of the northern summer, blind and helpless. In 10 days, their eyes open and they begin to walk. They seldom stray from the den for the first six or eight weeks of their lives.

Lynx mothers have been seen traveling together with their young playing as one happy litter. No one knows for certain if these mothers share their kills with each other's kittens, but it is assumed they do. The father is long gone, which is just as well for the kittens. He is likely to eat his own offspring if he encounters them. Why feline fathers offer no help in the rearing of their young is not clear. Kittens, it would seem, could just as easily benefit from a second parent bringing home game as wolf or fox pups do. Certainly, the task of hunting for herself and three or four rather large young until the following spring when she abandons her litter is a demanding one for a single mother lynx. It's even more taxing for cougars, which hunt for their young for up to two years. Perhaps it is simply not possible to switch on and off all the solitary and stealthy behaviors that make cats such efficient hunters—although the mother appears to manage it well enough with her kittens.

The lynx and the bobcat seem to have neatly partitioned the continent. While the lynx remains unseen by keeping to the boreal forest, north of most centers of human habitation, the bobcat lives among us. A map of its range looks like a slightly bloated map of the contiguous United States, spilling over its borders with Canada and Mexico. Remarkably, the bobcat not only survives throughout this range today, it thrives. There are between 700 000 and 1 500 000 living in the United States alone. (The cougar, which once roamed almost the whole continent south of the boreal forest, now numbers perhaps 20 000. It is confined almost entirely to a range including and west of the Rocky Mountains.)

The bobcat's success stems from its willingness to eat anything from trout to bats. Because of its appetite for smaller animals, it can hunt a much smaller territory than a cougar. It prowls the wild places between our cities and farms, in swamp, desert, woods, and scrubland—anywhere it can catch the rodents, rabbits, and hares that are the mainstay of its diet. Where food is plentiful, it is content with a territory as small as 2.5 square kilometres (one square mile). Where food is scarce, it may have to hunt a much larger area, up to 100 square kilometres (40 square miles) to feed itself. By comparison, a cougar's range may be five times a

LEFT: *Beguiled by their blue eyes and cuddly charm, people have tried to make pets of cougars over the years, and in many places it's legal. But one has only to imagine the consequences of even the most docile housecat's occasional fit of temper magnified about 15 times to realize the disastrous possibilities of raising a cougar in captivity.*

bobcat's maximum range, and lynxes are rarely content with less than 90 square kilometres (35 square miles) in which to hunt.

Despite their numbers, very few people will ever see a wild bobcat. Like the cougar, they move unseen, belly to the ground, before springing on their prey. Although it primarily hunts smaller animals, the bobcat will tackle game up to eight times its own weight. In the Everglades, bobcats are responsible for almost 17 percent of adult white-tailed deer deaths—remarkable when one considers the bobcat's size. Although one giant weighing 27 kilograms (60 pounds) was caught, most males are a third that weight. Females are somewhat smaller.

Bobcats kill with the same, single bite as the cougar and the lynx, delivered either to the spine or throat. A study in Florida's Big Cypress National Preserve in which deer were fitted with radio collars showed that although the bobcat usually took prey that had bedded down, some of the largest deer were killed on the hoof, often carrying the bobcat a considerable distance on their backs before succumbing. The deer were often found with their heads twisted under their bodies. This probably happened as the bobcat dragged its prey over the ground. However, some biologists believe that bobcats suffocate their prey by clamping their

ABOVE: *Cougars are capable of leaping across chasms of 9 metres (30 feet). This is just slightly more than the same distance covered by the current world record–long jumper. The performance gap would widen considerably, however, if the human and feline contestants competed with deer carcasses clamped in their jaws. A cougar can still bound 3 metres (10 feet) with ease, while the human would be unlikely to leave the ground.*

LEFT: *Lynxes will climb trees to escape their enemies, which are primarily humans, wolves, and coyotes. The only time they are vulnerable to their natural predators is when they are caught in the open, but these cats are extremely shy and rarely venture from the trees.*

mouths to the animal's neck and wrenching it backward to pinch the windpipe shut.

The bobcats then fed on the meat of the hindquarters, the lungs, the heart, and other organs. One of the difficulties of determining just how they kill their prey stems from the cats' habit of also consuming much of the flesh around the throat and neck, obliterating clues to the killing bite.

There is another reason that the bobcat has fared so much better than the cougar in the face of urban sprawl: The bobcat does not attack human beings. Bears are responsible for far more maulings each year than are cougars, and wolves kill more livestock. But for some reason—perhaps its unfortunate tendency to attack children more often than adults—human beings reserve a special ire for the cougar, and we have been relentless in exterminating it from much of its native range. Nor have we made any effort to restore cougars to their former haunts, as we have wolves and lynxes. In 1999, 11 lynxes were released in the San Juan Mountains of Colorado in a so-far unsuccessful effort to re-introduce the cats to the state for the first time since 1975. (It appears that the lynxes are having trouble finding enough hares to feed themselves.) But lynxes, like bobcats, pose no threat to people.

Of late, the public has come to admire the more benign attributes it recognizes in the society of the wolf pack. The bears, while probably more dangerous to an adult human than a cougar, we have always imbued with a certain loutish charm. But the cats are the purest of predators, and our admiration of them will probably always be tinged with fear. The invisibility that has so ably served them as predators now serves them again as prey.

RIGHT: *Apart from the fisher, cougars are one of the few predators that commonly kill porcupines. The cougar slips a paw under the porcupine, flips it over, and bites the stomach, the one place unprotected by quills. It's no job for an amateur: Young, inexperienced cougars have died from wounds inflicted by porcupine quills.*

Led by the Nose

The dogs—wolves, coyotes, and foxes—live in a world of smells that human beings can only imagine. We can measure the sensitivity of a dog's nose, which is some 100 times that of ours, but we can hardly know what smelling is like for a dog. The best we can do is to draw an analogy with our own most developed sense, vision. The difference between our sense of smell and a dog's is probably akin to the difference between glancing at a photograph of a room and standing in that room and being able to look all around it. One experience is flat and momentary; the other has the added dimensions of depth and time. Both experiences are informative, but one is immeasurably richer than the other.

Except when eating or presented with singular, strong odors, most of us are unconscious of our sense of smell. We think of substances or objects giving off "smells," as if an odor were something apart from the source. But when any animal—including a human being—smells something, it is actually taking in particles of that something and identifying them. A wolf can smell an elk just from the microscopic bits the elk has shed in walking by, perhaps weeks earlier.

There are other animals with even more sensitive noses than those of

ABOVE: *The swift fox has a patchy distribution throughout the western half of the United States and the extreme south of the Prairie provinces. Its preferred habitats are deserts and short-grass prairie, where flat-out speed is a great advantage in running down prey. The swift fox can hit 40 kilometres (25 miles) per hour in a sprint.*

LEFT: *As most of its prey are strong runners, the wolf's survival in turn depends on its ability to run. Wolves cannot match the top speed of a deer or a moose, but they may hound their prey in relays, chase them into ambush, or press other advantages of the pack to catch their quarry.*

the canids: Polar bears can track down a seal carcass from 20 kilometres (12 miles) away, and a black bear can distinguish the contents of unopened canned goods. But in canids, the sense of smell is uniquely bound with keen hearing, good vision, and intelligence to produce a formidable predator.

Wolves are arguably the smartest of North America's carnivores, and many biologists credit the evolution of their intelligence to the coordination of their hunting efforts. For any predator, finding and attacking prey involves many decisions: weighing effort against risk, choosing when to reveal oneself, guessing which way the prey will run. But in addition to these decisions, a wolf must choose a role in the hunt and monitor the actions of its fellow pack members. Securing food is not the only consideration. Some biologists even believe that wolves make political decisions on when to impress the pack by assuming risks during the hunt or deferring to other members during the division of the spoils.

It is the pack that makes wolves such effective hunters. They don't have the same equipment for killing as the cats. Their bite does not have the force of a cougar's, nor are their limbs and claws suited to grabbing and immobilizing prey so that they may deliver a precise, lethal bite. But their numbers, combined with stamina that will carry them at a steady 40 kilometres (25 miles) per hour for at least 20 minutes, enable them to hound animals as large as bison and moose to death.

After harrying and exhausting their quarry, the pack tears at it with multiple, lacerating bites that eventually bleed the prey into shock. Wolves may not be deliberately cruel, but one thing is certain: Their prey suffers a gruesome death.

Wolves do not always eat large game. At the northern extremes of their range, which covers Alaska, Canada, and a few patches in the northern contiguous states, hares and rodents form a significant part of their diet, as they do whenever hoofed animals are scarce. They may apply some of the same group strategies to flush and intercept these smaller animals that they use on larger prey.

The pack is a hierarchy. At the top are a dominant male and female called the alpha pair. Often they are the oldest and founding members of the pack. The alpha male or female initiates the hunt, howling, and other group behaviors. All other pack members defer to the alpha pair.

The alpha male is the first to feed on a kill. Unless the pack is colonizing new territory, only the alpha pair produces offspring. Whether the

ABOVE: *You can tell a wolf's position in the hierarchy of the pack by its posture. A wolf standing tall and erect, such as the one on the right in this photo, is dominant. The wolf approaching from the left with knees bent is deferring to the dominant wolf. Lying on the back with belly exposed is an act of supplication, and the most submissive position of all.*

RIGHT: *The red fox is a specialist at mousing, and it has several adaptations that give it a highly accurate pounce. Its ears can determine the direction of a mouse's rustling within one degree. By flailing its long, bushy tail, the red fox can actually change course in mid-air. Finally, its skeleton is especially light and strong. Foxes are only half the weight of a dog the same length.*

other pack members are prevented from mating through intimidation by the alpha wolves, or some physiological condition prevents subordinate females from coming into estrus, no one is certain. Female wolves enter estrus once a year, and the opportunity for successful mating lasts only five to seven days. During that time, the alpha male never strays more than a few steps away from his mate.

Subordinate female pack members are reluctant to mate even if males do attempt to mount them. They thwart copulation by sitting down or curling their tail between their legs. On some instinctive level, they seem to know that their offspring would be denied the full support of the pack—something that alpha pups are guaranteed. All pack members help to rear the pups, which are usually born in litters of six or seven. Adults will indulge them in play for hours on end, and defend them when danger threatens.

Nonetheless, the day comes when an alpha wolf's status is challenged by a younger rival. Sometimes, the challenge is a physical attack and the two wolves fight. But challenges may take other forms—attempting to mate with the alpha female, initiating pack activities, or assuming responsibilities normally taken by the alpha wolf.

For example, the alpha wolf will sometimes subdue large prey such as moose by clamping onto the moose's nose with its jaws. This makes it almost impossible for the moose to maneuver, and safer for the rest of the pack to attack. But it's a very risky tactic for the wolf that takes the clothespin position, for it is within reach of the moose's lethal front hooves. Still, a wolf seeking to elevate its status in the pack might elect to take the risk.

If there is an alpha, there must also be an omega, and most packs include a wolf of low status, usually a young female, suffering at the bottom of the hierarchy. Sometimes this animal endures the frustrations and abuse passed down the chain from the alpha pair. It may eventually choose to start a new pack or forego the advantages of pack life completely and become a loner.

Moment to moment, each wolf either holds or seeks a place in the hierarchy through expressions and posture. Cringing, lowering the tail or head, and wagging the tail all signify submission. Rolling over on the back is the most submissive position of all. Staring directly at another wolf, growling, and mounting are all efforts to establish dominance or superiority. Like domestic dogs, wolves also convey their intentions through a variety of facial expressions.

And so what may have begun as a mated pair combining their hunting efforts in the distant past has evolved into a society, with some very complex behaviors to keep it all functioning. But there are costs to pack life. Avoiding the abuse of your cohorts or dishing it out to them takes vigilance and effort. The overall caloric demands of the pack are high, even though little meat goes to waste when it takes down a big animal. A pack may be anywhere from 2 to 20 individuals, and canids do not have very efficient digestive systems. Particularly in lean times, the pack has to

RIGHT: *This wolf pup is only three weeks old, but already it knows how to howl. Howling is instinctive among canids. Sometimes it can be stimulated by any protracted, falling note, as when domestic dogs are set to baying at a passing siren. At other times, however, even high-fidelity tape recordings have failed to induce wild wolves to join in.*

ABOVE: *At three weeks old, these wolf pups are ready to leave the den, although they will not wander far from it for another five or six weeks. During this time they will remain within sight of their mother or another pack member. By their first autumn, these pups must be strong enough to travel with a pack that may cover 20 kilometres (12 miles) in a day.*

RIGHT: *The gray fox is the oldest of the North American canids, having evolved some 6 to 9 million years ago, and the only one able to climb trees. Although this skill allows it to forage for nuts or orchard fruits, climbing may be more useful for escaping its natural enemies. Historically, these were wolves, but today coyotes, domestic dogs, and humans are the main predators of foxes.*

range over a much larger area to feed itself than a lone wolf would. Although five wolves can combine their abilities to kill, travel costs can't be shared; each individual will still have to run over five times as much territory to find enough to feed them all. A wolf pack's territory may take in 650 square kilometres (250 square miles) where game is sparse.

Within the pack, the hierarchy is maintained by posture, staring, growls, facial expression, and other signals of dominance or submission. Rarely do wolves fight with fellow pack members. Disputes between packs, on the other hand, can take a heavy toll. Two wolf packs encountering one another usually fight, sometimes to the point where wolves are killed.

There is a system of safeguards to prevent this eventuality. Pack members establish the boundaries of their territory by marking them with urine and scat. They howl to let adjacent packs know where they are. The individual members howl in different keys, to maintain separate voices. In this way, the pack sounds as populous as possible. If they were to howl in perfect unison, a rival pack might believe the neighboring pack to be smaller than it really is and so be emboldened to challenge it. It's all part of an elaborate system to keep packs sharing territorial borders from ever meeting.

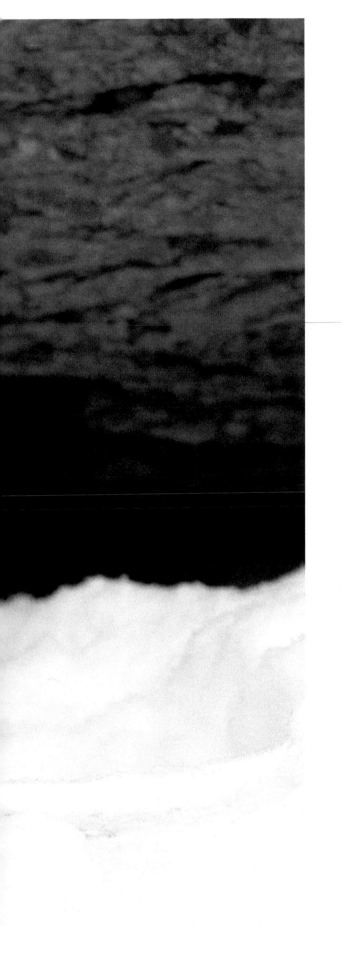

Unlike wolves, coyotes are not bound to the pack by the need to kill large game. They eat mainly rodents and rabbits supplemented by a variety of other foods: fish, crayfish, fruit, frogs, lizards, bats, beavers, deer, house cats, small dogs, and even insects. In short, whatever they can find. There's little advantage to being in a pack when hunting such small prey. Coyotes usually travel in pairs, but often singly. Those that do live in packs generally feed on carrion, a situation in which defending a kill may require numbers.

While both the wolf and the coyote have been the targets of human campaigns to eradicate them, the coyote has gradually increased its range. At the beginning of this century it was confined primarily to grasslands and deserts at the edge of the gray wolf's territory. But as the wolf—the coyote's major competitor—has steadily been eliminated from the contiguous United States, the coyote has spread outward to both coasts. Now it lives in all but the southeastern corner of the United States, as far north as Alaska in the west and James Bay in the east.

The coyote thrives amid urban sprawl because of its willingness to eat almost anything. It lives in some of our largest cities. Many city dwellers, while fearing for Fluffy's safety, bear a grudging admiration for the coyote's tenacity. They might as well, because there is little they can do about this cunning interloper. Even if the coyote were not so adept at avoiding traps and poisoned baits, the residual kill of domestic cats and dogs would bring a quick halt to any urban campaign against it. Like the cockroach and the rat, the coyote has found a haven beneath the table of its persecutors.

Foxes have also benefited from the wolf's near extermination in the contiguous United States. Some biologists believe that today's North American red fox is in fact a hybrid of a relatively rare strain that was

LEFT: *The kit and swift foxes were once considered separate animals but are now recognized as a single species. Cropland replaced much of this fox's habitat and poisoned baits intended for coyotes and wolves killed so many of them that they were nearly wiped out by the 1930s. Once eliminated completely from Canada, they have been re-introduced through breeding programs.*

RIGHT: *Like the coyote and the bobcat, the red fox thrives in the spaces between farmlands and cities, feeding mainly on small mammals and birds. The red fox also eats fruit, acorns, insects, reptiles, and amphibians. With the virtual collapse of the fur trade and chicken ranching now almost an industrial enterprise, there are few reasons for human beings to shoot foxes, but thousands are killed by automobiles every year.*

native to North America. It may have bred with the European red fox when it was introduced to the New World as recently as 1850 by English immigrants wanting to hunt their old quarry. Whether this is true or not, the red fox has expanded its range to the point where it is the world's most widespread carnivore. In North America, it lives from Georgia to Baffin Island, in almost any kind of habitat except for coastal temperate rainforest and the Everglades. The gray fox's range is smaller, covering all but the Northwest quadrant of the contiguous United States.

The red and gray foxes are more omnivorous than other wild dogs, and though they prefer to eat small mammals, birds, and eggs, they will readily switch to a diet of fruit and nuts when they are in season. So there may be some truth to the fable of the fox and the grapes, after all. The gray fox even climbs trees—though probably more to evade coyotes or other predators than in pursuit of fruit. Foxes supplement their diets with reptiles, amphibians, and insects.

Unlike wolves, foxes are solitary outside the breeding season, but they share the canid intelligence. A red fox was seen charging at a foraging ground squirrel. The squirrel retreated to its burrow, but after several minutes returned through a back entrance to the spot where it had been feeding. The fox pounced again, and again the ground squirrel escaped, later to re-emerge from the back door. The third time the ground squirrel ran into its burrow, the fox went immediately to the rear entrance where it waited with its mouth open. When the ground squirrel emerged, the fox snapped it up.

Like all the canids, foxes raise their young in dens, although the fox is more likely to enlarge a badger or a groundhog den than it is to start from scratch. The male helps with rearing the litter of five to seven pups. When the pups are about two months old, they are weaned and their parents start to bring them rodents they have crippled. The young foxes learn how to chase and catch their prey by playing with the injured animals. Later, the parents take their pups hunting. By the fall of their first year, the young foxes strike out on their own.

The swift fox, a cat-sized relative of the red fox, feeds mainly on rabbits and other small mammals, and is named for its ability to chase down its prey. It's been clocked at 60 kilometres (37 miles) per hour. It was once common on the Canadian prairies, the deserts of the great plains of the United States, and Mexico.

Of all the dogs, perhaps the most impressive survivor is the arctic fox.

It thrives in the open in one of the coldest climates on earth, a remarkable feat for an animal barely larger than a house cat. It's protected from the cold by its luxuriant fur coat and the ability to increase its metabolic rate when the weather drops below –60°C (–75°F). As with many animals living in very cold places, its ears and snout are blunter than those of other foxes. The arctic fox's muzzle has been shortened into a cone. These extremities are also kept from freezing by venous heat exchangers. The arteries supplying the feet lie alongside the veins carrying blood away from them. Arterial blood entering the feet loses most of its heat to the cold, venous blood returning to the body before it ever enters the feet. In this way, enough heat enters the feet to keep them from freezing, but not so much that the fox loses excessive heat to the environment. On really cold days, when the arctic wind blows across the tundra like an ice scraper, the fox hunkers down with its tail wrapped around its face like a scarf.

The arctic fox feeds mainly on voles and lemmings, and may eat a thousand of them in a summer. In winter these rodents forage beneath a blanket of snow, but the fox's sharp nose enables it to dig them out.

Arctic rodent populations are often cyclic, however, and when their numbers drop, the fox ventures away from solid ground and out onto the pack ice to seek other food. There, it plays a dangerous game of tracking a polar bear, sometimes for weeks, feeding off the remains of its kills. The fox is usually too agile for the polar bear to catch, but the bears are experts at ambush. If it detects its pursuer, a polar bear may huddle downwind of the trailing fox as it would at a seal hole and simply wait for the fox to blunder into it.

The canids have taken a forked evolutionary path since the arrival of modern hominids. The wolves human beings adopted eventually became the domestic dog in all its forms. Under the eaves of our buildings and on the flagstones of our hearths they have become the most successful and diverse of the canids, far outnumbering their wild counterparts. Many live long and pampered lives, with special diets and medical care lavished upon them. In North America, few dogs have to work for their suppers.

For the wolves that had the temerity to remain wolves, we reserved a particular dread, portraying them for centuries as bloodthirsty killers. This isn't too surprising, considering that early human hunters competed

BELOW: *Coyotes can run faster than wolves, sprinting up to 65 kilometres (40 miles) per hour. At such speeds, they can chase down hares, a staple of their diet throughout much of their range. Coyotes are also great travelers, and tagged animals have been recovered more than 640 kilometres (400 miles) from their point of capture.*

OVERLEAF: *Wolves have large feet compared to dogs, and by splaying their toes they can walk on crusted snow without breaking through. Fresh snow will not support them, however, and when it is more than chest deep their longer-legged prey such as moose, deer, and elk have a definite advantage.*

ABOVE: *This coyote is mousing, alert for any rustling that will betray the location of its prey before pouncing. Wolves live only where there is large game for them to hunt at least part of the year, but coyotes can survive happily on prey no larger than a rabbit.*

RIGHT: *In summer, the arctic fox will hunt past the point of satiation, then dig down to the permafrost layer to cache the excess in a natural deep freeze. Lemmings are an important food for these foxes, whose populations rise and fall with the rodents' four-year cycle. In winter, many take to the pack ice where they follow polar bears, feeding on the remains of their kills.*

directly with wolves for game, and that ranchers also saw them as a threat to their livelihood.

The foxes and coyotes fell somewhere in between. Some have found homes in our cities, but many more have been poisoned or trapped. They have been persecuted for being one of the main carriers of rabies. They, too, have been hunted for their fur.

Of late, the wolf's image has undergone a dramatic reversal. Our discovery that they have rich social lives and that the cohesion of the pack depends upon courage, loyalty, and even respect as much as it does hunger has led us to re-evaluate our image of the wolf. Outside of whales and dolphins, it may now be the most loved of all wild animals.

Perhaps we would have seen those qualities sooner in the wolf had we not attributed them exclusively to our dogs for so many centuries. Or perhaps without the dog, we never would have recognized them at all. Undoubtedly, domestication has at once confounded and enriched our special relation to the wild dogs.

Hobos

It takes a lot of landscape to dwarf a grizzly bear, but the Cambria Ice-field is equal to the task: From its center, just east of the southern tip of the Alaska Panhandle, ice, snow, and very little else stretch to the horizon in every direction. There's nothing to eat and no reason for a bear, indeed any animal with two cells to rub together, to be here. But here it is. The grizzly rises to its hind legs to examine the buzzing helicopter that circles for a better look at it, then drops to all fours again and shambles on.

The pilot shakes his head. "Those animals are the damndest walkers."

If the bear could respond to the pilot, it might well say, "And what's to stop me?" The answer is, nothing. The grizzly bear has no natural enemies. But it seems curious behavior nonetheless. While most animals are scrapping for every calorie, bears sometimes act as if they were on a weight-loss program. When they're not feeding, they're usually walking.

It's not as though the bear's body is particularly suited to its perpetual march. It does not have the slender legs of a caribou or a dog that make for efficient long-distance walking and running. Bears, like people, are plantigrade: They walk on the whole soles of their feet, and that slumping gait costs a great deal of energy. Nor would looking at its shaggy hams

ABOVE: *The rainforests of the north Pacific coast are highly productive, but most of what they produce, the wood and needles of coniferous trees, is inedible to all but a few mammals. As a result, the temperate rainforest is almost devoid of large animals. But Pacific salmon runs bring nutrients from the sea to this black bear.*

LEFT: *Near Churchill, on the shore of Hudson Bay, up to 120 bears gather to wait for the sea ice to form before walking to their winter sealing grounds. Here, adult males often wrestle playfully. One theory is that they are gauging each other's strength as a prelude to spring, when they may have to fight in earnest for access to sows. Another is that, well-fed and bored, the males simply have time and energy to spare.*

lead one to believe that a bear can outsprint a horse at full gallop. A relative newcomer to North America that crossed the land bridge from Asia just 30 000 years ago, the grizzly, or brown bear as the Alaska coastal variety is called, seems determined to check out every nook of the continent before deciding whether or not to stay.

A more likely explanation for the restless behavior of both grizzlies and black bears is to be found in their diets. At various times of year, these omnivores eat almost anything that a mammal could consider food: grasses, roots, sedges, berries, salmon, fungi, moss, mice, marmots, ground squirrels, turtles' and birds' eggs, cereal crops, cattle, carrion, and the young of elk, moose, deer, and caribou. They even eat insects, using their powerful forepaws to tear open the nests of ants, bees, termites and their grubs. Grizzlies and black bears will sometimes dig for hours to catch rodents they have trapped in their dens.

With such a varied menu, almost any direction is a path to riches. And so they walk. Bears are big animals; the polar bear and the grizzly are the largest of all land carnivores. Males of either of these species grow to 550 kilograms (1200 pounds). To support such bulk they must patrol and occasionally defend large home ranges. One study found that male grizzlies (which have territories three or four times the size of a female's) in the Brooks Range of Alaska regularly wandered over 1350 square kilometres (520 square miles), an area considerably larger than greater Washington, D.C. Grizzly bears depend heavily on seasonal foods: berry crops, salmon runs, or caribou migrations—depending upon where they live. In a given year, some of these harvests will be abundant and others not. A bear with a large home range has a greater diversity of food sources from which to choose.

The polar bear, which evolved from the grizzly during the last ice age,

RIGHT: *At the McNeil River State Game Sanctuary in Alaska, scores of grizzly bears congregate to fish for spawning chum salmon. Because there is abundant food, for a few weeks the rules of territoriality are relaxed and bears will tolerate one another at close quarters. Even the human tourists go largely ignored.*

has its own reasons for wandering. Polar bears eat marine mammals—mainly seals, but also walruses and even beluga whales. They all require access to air through openings in the sea ice where the polar bear hunts. The locations of these openings change from day to day on the drifting pack ice, and the polar bear may have to walk a long way to reach its next meal. One male polar bear, tagged for a study, crossed the entire Arctic. It took a year, but the bear walked 3200 kilometres (2000 miles).

The pads of the polar bear's paws are hairy, which keeps them from freezing or slipping on the ice during its long journeys. Its front toes are partially webbed. Although all bears are good swimmers, polar bears are particularly fast. They swim at speeds up to 10 kilometres (6 miles) per hour using only their front paws and are so comfortable in the water that some biologists classify them as marine mammals. They have been found out to sea at distances that would have taken them hours to swim. Although they will sometimes dive for food, the ability to swim is more important to these bears because they spend so much of their lives on the sea ice. During spring breakup, they have to swim between ice floes. Animal navigation in general is one of the enduring mysteries of ethology, but polar bears manage to navigate in an environment where the

ABOVE: *Polar bears are the most carnivorous of the bears. In summer they eat a little moss and some sea weeds—perhaps to obtain certain vitamins—but during winter they feed almost exclusively on seals. On one occasion, a polar bear was seen diving onto a beluga whale that kept returning to a shrinking breathing hole in the sea ice.*

LEFT: *The grizzly bear is a newcomer to North America. Only 30 000 years ago, it crossed the land bridge joining Alaska and Asia and spread southward, displacing black bears as it went. Today, there may be 50 000 living in Alaska and western Canada. Twice that number still inhabit the forests and tundra of northern Asia, and a remnant population of about 60 hangs on in the mountains of Spain and Italy.*

sky is often clouded over and the landmarks are continually shifting.

Polar bears are skilled and patient hunters. They usually catch seals by lying motionless at their breathing holes, sometimes for as long as two hours. If the bear moves, a seal may hear it and choose an alternative breathing hole. A story persists that polar bears cover their noses—the only contrasting parts of their bodies—with their paws to remain less conspicuous when ice fishing. Unfortunately, no one seems to have actually witnessed this phenomenon.

Polar bears have longer necks and legs than grizzlies, probably to extend their hunting reach. When a seal surfaces for what will probably be its last breath, the bear gaffs it with 15-centimetre (6-inch) claws. Polar bears are so strong that they can easily haul an adult ringed seal weighing 115 kilograms (250 pounds) onto the ice with one paw. Immediately, the bear drags the seal away from the hole to prevent any chance of escape and bites through the seal's thin skull. It then flenses the carcass of its body fat. When seals are plentiful, polar bears may eat just the fat before abandoning the rest of the carcass to scavengers such as ravens or arctic foxes.

Polar bears may also stalk seals. Although they can't hope to catch one in open water, they have learned that ringed seals dig lairs into the snow to bear their young. A bear can smell them through the roofs of their lairs, which may be a metre (3 feet) thick. It crashes through the roof by diving on it—more than once if necessary.

The grizzly also looks underground for much of its food. One of its distinguishing characteristics, the shoulder hump, is a mass of muscles that gives those forepaws the power to churn sod. Like the polar bear, it has huge claws. It uses them to unearth roots, tear apart stumps to reach insects, and dig out one of its favorite foods, the ground squirrels. These rodents are an important food in late summer, when they have fattened

LEFT: *A grizzly bear awaits the arrival of salmon fighting their way up a waterfall. To a fish, an immobile grizzly probably appears little different from a rock, particularly in the turbulent, air-filled water at the base of these falls. A bear's best strategy under such conditions is to simply wait with open jaws for a leaping salmon and snatch it from the air.*

up to prepare for their hibernation. At that time of year, the grizzly has much the same thing on its mind, and actively hunts these squirrels.

On sighting a bear, a ground squirrel gives a warning whistle and retreats to its hole. In pursuing its prey, the grizzly has been known to make an excavation so large that it may disappear underground itself. Occasionally, the bear lifts its head to make sure that the ground squirrel doesn't escape out a back door. If it bolts, the grizzly will give chase and may even dig it out of a second den.

On the coasts of British Columbia and Alaska, the preferred food of all bears is spawning salmon. Grizzlies and black bears descend upon runs of coho, chinook, sockeye, pink, and chum to practice a range of fishing techniques. Sometimes they will splash through the shallows until a salmon reveals itself by thrashing through a riffle. In deeper water, they search with their heads submerged, periscope fashion, where refraction probably improves their myopic vision. Others sit like a rock in the middle of the watercourse until the salmon forget it's there. When a salmon blunders into the bear, the bear pounces to its elbows, clamping it between its paws, or scoops it out of the water and onto the shore in one motion. At the famous Brooks Falls on the Katmai River in Alaska, grizzlies will station themselves at the top of the falls with jaws agape and snatch leaping fish from midair.

Once a bear catches a salmon, it usually moves away some distance to eat it. Bears are not social animals, but where many of them are drawn to a food so plentiful that there is enough for all, such as a salmon run, they generally tolerate each other's presence. Except for the mothers with cubs, most of them are there alone, doing their best to avoid each other and concentrate on their fishing. Of course, there are dining protocols. The largest bears, usually boars, stake out the best fishing spots. Older mothers with cubs are in the second rank.

Although smaller than the largest males, experienced mothers make up what they lack in weight with aggression. First-time mothers may be distracted and skittish, on the lookout for boars threatening their cubs. Males will often take the opportunity to kill cubs that aren't actively defended by their mothers. At the bottom of the social hierarchy are the two- and three-year-olds, often sparring and mock fighting when not fishing.

During spring and early summer, grizzlies also hunt the young of caribou, mountain sheep, deer, musk oxen, and other hoofed animals. Most adult ungulates are too fast for the grizzly bear, but often their newborns

are not. Musk oxen, bison, moose, and elk will almost always try to fend off a grizzly attacking their young. Sometimes they are successful.

Seasonal game differentiates the diet of grizzly bears from that of their relatives the black bears, which arrived in North America several million years before their larger cousins. Black bears are far more likely to browse or fish than they are to hunt big game.

Grizzly bears inhabit more open spaces than black bears. They require some alpine meadow or tundra and once lived right across the prairie. Black bears are capable of foraging in the open, but as the larger and more aggressive grizzly spread down through North America, it probably drove the black bears of the open range into forests, where they could escape the grizzly by climbing trees. When black and grizzly bears meet, the black bear will either retreat or climb out of trouble. Due to their weight, grizzlies are rather poor climbers.

The grizzly may be more aggressive precisely because it evolved on the open plains. If a young grizzly were threatened—usually by another, adult grizzly—it had nowhere to hide. It, or its mother, would have to fight. Natural selection favored the more aggressive animals. Black bears, in contrast, have always had the option of retreat when dealing with danger.

Grizzlies would seem to have been the victors in the battle for territory, but that changed with the arrival of Europeans. Once, brown bears roamed all of North America. Today, they are confined to the western mountain chains, Alaska, and the tundra west of Hudson Bay. On the other hand, almost any wild area on the continent still supports black bears, even though hunters and conservation officers shoot tens of thousands of them every year. They have better withstood human encroachment mainly because their territories are smaller than those of brown bears. A black bear living in Tennessee may need only 9 square kilometres (3 1/2 square miles). The bear itself is smaller, with males weighing up to 270 kilograms (600 pounds) and females weighing half as much. Its diet is also more varied: In addition to the plants and small animals that grizzlies eat, many black bears will also eat human garbage. In the Okefenokee Swamp, black bears even raid alligator nests. Surveillance has shown that the alligators, normally fiercely protective of their eggs, retreat immediately when confronted by a bear.

Another reason for the black bear's ubiquity is that it does not require

BELOW: *Like most carnivores, bears copulate for a long time—sometimes up to an hour. Lengthy copulation may be necessary to stimulate ovulation in the female. While mammals such as deer or groundhogs must minimize any activity that interferes with their perpetual vigil against predators, a grizzly bear, having no natural enemies, has little reason to hurry.*

seasonal game, which has been eliminated from much of North America. Finally, as many a camper has learned with thudding heart, black bears do much of their foraging at night. It appears that bears living near people are more likely to be active in the dark.

Wherever winter snowfalls are heavy, the grizzly's and black bear's strategy of eating everything and anything begins to falter. The birds' eggs, the berries, and the termites are all gone. Bears are not skilled enough predators to hunt through the winter like a weasel or a lynx; nor are their digestive systems efficient enough for them to paw through snow to find grasses or other low forage in the manner of the ruminants. Although bears have a catholic diet, they select rather rich foods, and these simply aren't available to them in winter throughout much of North America.

Polar bears do not face this problem. Their main food, seals, are abundant year round. In fact, for polar bears at the south of Hudson Bay, summer, when the seals move north with the retreating sea ice, is their leanest time.

After the summer of gorging themselves to build their fat reserves, most black and grizzly bears seek a den in which to pass the winter. It may be a natural cave, a hollow tree, or a hole entirely of the bear's digging. An existing cavity is preferred, but these are seldom of the appropriate size. Ideally, the space should be just large enough for the bear to squeeze through to minimize exposure to the elements. The bear may collect leaves, twigs, moss, grass—any vegetation it can find—to line the den. Black bears prefer tree cavities high above the ground. These are more common in the southern states, where the hardwood trees grow that most often develop large cavities.

But why should a bear living in the mild climates of the south have to seek a winter den? In fact, there is no reason for a male bear to do so, and

LEFT: *The older a bear gets, the less inclined it is to play, but at the same time most bears retain a natural curiosity about their environment that stays with them for life. This grizzly began pulling repeatedly on this snow-covered spruce—a plant that it does not normally eat and would appear to have no interest in except as an object of play.*

boars living in mild climates stay active through the winter. But hibernation is more than just a way to conserve energy for the bear. It is also the time when females bear their young—a reproductive strategy unique among mammals.

Polar bear sows dig a cave into a snowdrift in November and bed down for four months to give birth to and suckle their cubs. At the southern extreme of their range, around James Bay, they dig earthen burrows into the banks of a stream or river. It would seem that this is a strategy to protect the cubs both from male polar bears and extremes of cold during the first and most vulnerable months of their lives. The males may not hibernate at all, and if they do, it's for a much shorter period.

For a bear, hibernation is not the profound torpor of a bat or a marmot. During hibernation, the body temperatures of these small animals may drop to only a few degrees above freezing, their hearts may slow to 20 or fewer beats per minute, and outwardly, they appear dead. Because bears are large, well furred, and extremely good at accumulating fat over the summer, they can doze through the winter with their bodies just a few degrees below their active body temperature and emerge in the spring with calories to burn.

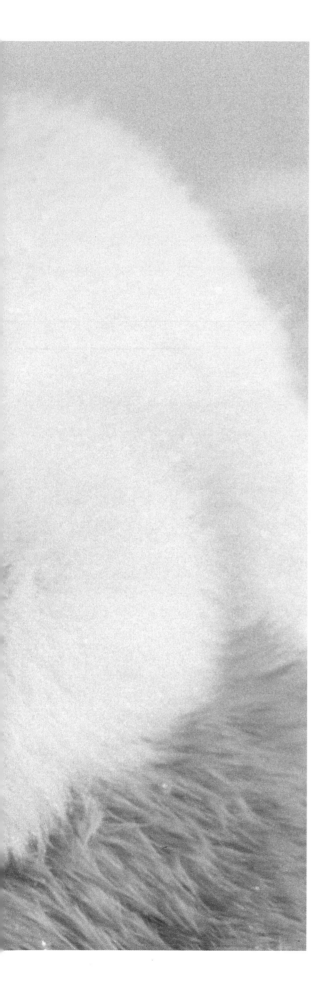

Born in the dead of winter, the bear cubs pass their first and most precarious months in the warmth and protection of the den. They emerge in the spring much larger and stronger than they were at birth, and by the end of the summer, they are better able to withstand their first winter.

Hibernating bears don't just collapse into a deep sleep. For a few weeks before they bed down for the winter, they remain close to their chosen den site, sleeping more and becoming increasingly lethargic during their waking hours. Once the bear enters the den for the last time that year, its heartbeat drops from 50 or 60 beats a minute to about 10. Its body temperature falls slightly, and its metabolic rate slows to about half. Blood flow to the limbs is restricted.

The depth of the bear's torpor depends on a variety of factors, but in general the colder it is, the more deeply it sleeps. In South Carolina or Florida, a hibernating bear may awaken and lift its head at any sound made by an intruder. A bear in Quebec might have to be poked or shaken for a few minutes to elicit any reaction. Once roused, it could be several minutes before it can walk. But there are many exceptions to this pattern. Researchers in Alaska tracked grizzly bears wearing special radio collars to their hibernacula. The rate of beeping from the researchers' receiver was proportionate to the rate of movement of the collar. Thus, the rhythm of the beeping was at its slowest when the bear was completely still. Biologists approaching the den of hibernating brown bears found that the beeping often accelerated when they were as far away as 40 metres (130 feet) from the den. Presumably, the bears heard the people's footsteps on the snow.

Although its sleep is much more profound, the bat or the marmot must waken periodically to urinate, defecate, and to drink. Rousing itself in this way is very costly, and burns much of the animal's winter fat. The

LEFT: *After emerging from their winter den, a mother polar bear and her cubs soon begin walking toward the edge of the sea ice to hunt for seals. This may be a trek of 16 kilometres (10 miles) or more. If she has to swim through open water on the journey, she will often carry the cubs on her back. Cubs still do not have enough fat to protect them from the Arctic Ocean's chill.*

bear, on the other hand, can pass the entire winter—up to six months—without drinking, defecating, or urinating. This physiological feat is unmatched by any other mammal.

Most mammals living off their fat reserves must also burn some protein to meet their metabolic needs. One of the products of protein metabolism, urea, accumulates in the blood unless it is excreted in the animal's urine. Urea is a toxin, and if an animal doesn't urinate it will eventually poison itself. Bears have a different way of eliminating urea from their blood. They use it to build new proteins. In this way, the bear conserves water and keeps urea from building to dangerous blood levels.

The female bear actually does eat and drink a little during hibernation. When a typical litter of one to three cubs is born, she licks them clean of amniotic fluid and eats the placenta. At 600 to 700 grams (21 to 25 ounces), the cubs are among the smallest of all mammalian babies in relation to their parents' size. With their hairless bodies and eyes sealed shut, they look like naked squirrels. The bear's gestation period is short for animals their size—from six to eight and a half months, depending on the species and the period of delayed implantation. They are, in effect, premature births. Nursing a cub takes more of the mother's energy than would keeping them in her womb, so why it is that the cubs are born so early in their development is something of a mystery.

When the mother curls into a ball, the nipples protrude into a confined airspace, warmed by her breath and body heat. Here, the cubs can feed in comfort. As they nurse, they make a contented humming. When they're hungry, they bawl in a loud voice. After nursing, the mother rouses herself and stimulates her cubs to defecate by licking their rumps. She eats their feces and urine to keep the den site clean.

After three to four months of nursing, the bear family emerges from the den. At this time, young black bears weigh 3 or 4 kilograms (6 to 9 pounds); polar bear cubs weigh 9 to 11 kilograms (20 to 25 pounds). Their mother will have lost much more than her young have collectively gained: a quarter to a third of her body weight. For a few days she is lethargic and moves slowly. In contrast, the cubs emerge restless and playful, with a world to explore. Caution prevails, however, and brown and black bears remain in the area of the den for some weeks, often returning at night to sleep. Black bear mothers sometimes build daybeds near the den entrance, ready to send their cubs up a tree when danger threatens.

Within a month of leaving the den, the cubs, in the company of their

ABOVE: *Because of their size, complete lack of natural enemies, and a diet that consists largely of meat, some naturalists believe that the polar bear is the single most dangerous animal a person can encounter in the wild. Although attacks on humans by polar bears are much rarer than attacks by other bears, this is probably because so few people live within the polar bear's range.*

mother, begin their life of walking. In April and May, food is often scarce and brown and black bear adults may lose weight. The cubs, with their mother's milk to nourish them, gain weight steadily. They will stay with her for up to two years, denning together for their first winter.

A few weeks after emerging from their snow dens, polar bears begin the trek to the sealing grounds. All winter, the ice shelf has been building and the bears may have to walk 100 kilometres (60 miles) to reach open water.

For all cubs, their first year will be the most dangerous of their lives, which can be as long as 25 years. A third of black bear cubs, 40 percent of brown bears, and 50 percent of polar bears die before their first birthday. Wolves and coyotes may kill cubs that wander from their mothers, but by far the most common predators are male bears, many of which kill cubs on sight. Grizzly bears occasionally kill adult black bears, and have even been known to dig early hibernators from their dens and eat them.

For adult grizzly bears, human beings are their only enemies. In turn, bears are the only wild animals that people venturing into the wilderness in Canada or the United States have real reason to fear. To be sure, there are other dangerous animals in the woods, but a pair of heavy boots will protect against a snake bite and attacks by cougars on hikers are very rare.

Wearing a bell while hiking, storing food away from tents, and not running are all sensible precautions. But beyond these basic measures, the experts seem unable to agree completely on protocol around bears. After all of the foolish provocations by people have been eliminated, there still remain a small number of wanton attacks.

It is in our nature to assign rules and patterns to what we observe in nature, and our efforts become somewhat desperate with regard to animals that are dangerous to us. But we may never be able to predict what every bear will do, any more than we can consistently predict the reactions of individual people encountering bears.

We do know one thing: Almost all bear/human encounters resulting in human injury end with the bear being hunted down and shot. If, several thousand years from now, we have left any space for the bear, we may have selected against aggressive individuals so consistently that all bears may flee humans, or at least react more mildly than they do today. In effect, we'll have created the Teddy we've always imagined.

BELOW: *The grizzly bear is an omnivore that eats a variety of calorie-rich foods, from berries to elk. This shot gives a good look at its powerful yet flexible five-toed paws that may kill a deer with one swat or tear open a log to find the termites within. This grizzly is digging for razor clams.*

OVERLEAF: *Names such as Teddy bear and* Ursus horribilis *(an outdated scientific name for the North American grizzly) exemplify our ambivalence toward bears. The reasons for our fears are obvious. Why we find bears so beguiling is harder to say. Their broad faces and blunt features may evoke some of the same parental feelings that endear human babies to us.*

Overrun by Hooves

There are many defenses against the wild predators of this world, but few are as effective as being big. Although it does not protect against tuberculosis or tapeworms, size is a major deterrent to wolves and cougars. Large animals are not only better fighters, it's harder for a predator to get its jaws around a wide rump or a thick neck. They also win through intimidation. Most predators are reluctant to attack anything much larger or stronger than themselves for fear of injury, and so very large animals seldom *have* to fight.

Even if an animal lives by fleeing rather than fighting, size is still a help. Running speed is a product of stride length and frequency. In general, bigger animals have longer legs, take longer strides, and so run faster.

North America's biggest and fastest land animals are hoofed animals, or ungulates. This is true not just of the group as a whole; several are individual record holders. The bison, or buffalo, is the continent's largest land animal, with bulls reaching weights of nearly 900 kilograms (one ton). Next to the cheetah, the pronghorn is the fastest mammal in the world, having been clocked at 98 kilometres (61 miles) per hour. Caribou are the world's most efficient walkers. They move a given weight

ABOVE: *Moose are woodland browsers using their flexible lips to strip twigs, leaves, and bark from shrubs and trees. In fact, the name moose comes from an Algonquin word meaning "cuts or trims smooth." In summer, moose also do much of their feeding in the water, wading and even diving in shallow ponds or lakes for aquatic plants.*

LEFT: *The desert bighorn is a subspecies of the bighorn sheep and inhabits the cliffs and mountain ranges of Nevada, California, northern Texas, and Mexico. Its leaner, smaller body helps it to keep cool. Desert bighorns obtain much of their water from cactuses and other desert plants, and they may go days between drinks of water.*

of flesh a given distance using fewer calories than any other land animal.

The hoofed animals achieve their impressive bulk by eating plants, and only plants. The deer family, which includes the moose, elk, true deer and caribou, are mainly woodland browsers. Mountain sheep and goats, bison, and musk oxen graze for a living. Feeding on plants requires little energy or thought, but because grasses and leaves are so much less nutritious than seeds or meat, ungulates must eat a lot to grow and to maintain their weight. They have long and complex digestive tracts for extracting every available calorie from their food. With the exception of the collared peccary, all of North America's wild hoofed animals are ruminants and have four stomachs. Initially, they swallow vegetable matter with very little chewing. The grass or leaves are stored in the first stomach, or rumen, where saliva, enzymes, and bacteria begin to break down cellulose and other hard-to-digest structural components of the plants. Some of the food is passed along to subsequent stomachs, but the bulk of it is regurgitated hours later as cud, a ball of semi-digested plant matter that is chewed at leisure, often while the animal is resting, and swallowed again before being distributed to the other stomachs. In the first and second stomachs, enzymes and bacteria extract different nutrients from the food. The heat of fermentation even provides a little extra warmth for ruminants during the winter. The powerful acids responsible for digestion in omnivores are confined primarily to the fourth stomach, where many of the micro-organisms that aided in the earlier stages of digestion are themselves broken down, providing more calories for the ruminant.

Having bodies so effective at both deterring predators and feeding, one might expect the world to be overrun by hoofed animals. And if the reports of the first European explorers of North America are to be believed, once they did. Historically, bison herds on the grasslands of

RIGHT: *A white-tailed buck flashes its signature tail. This is thought by some biologists to present an easy flag for fawns to follow when their mothers enter dense brush to flee predators. Few hunters can keep up with the white-tailed deer in such circumstances. Adults can bound over a fence 2.5 metres (8 1/2 feet) high.*

North America were a force of almost geologic scale. There are accounts of them covering the landscape to the horizon in all directions, with more fur than grass visible. Some herds may have comprised as many as 4 million bison. A stampede was an earthquake on cloven hooves.

Bison probably reached their highest numbers some time after the arrival of the first wave of white settlers in North America. The effects of introduced disease and the war upon Native Indians reduced the numbers of the bison's chief predator so that the giant flourished. But it was soon realized that the best way to destroy the Natives was to eliminate their food source, and white hunters slaughtered bison by the millions for their hides and to feed the workers building the continental railways. The introduction of firearms also made the remaining North American Natives more effective hunters and bison populations plummeted.

Today, the bison is an icon of abundance lost. From their zenith, estimated conservatively at 60 million animals in the 17th century, bison numbers may have fallen to as few as 400 individuals by the 1890s. The largest wild herd lived in what is today Wood Buffalo National Park, straddling the border of Alberta and the Northwest Territories. The only other wild herd was in Yellowstone National Park, where poachers reduced their numbers to 23 individuals. The eventual efforts to save the bison from extinction resulted in one of the first successful efforts at wildlife conservation. Today, there are about 200 000 bison in both wild and managed herds, and their numbers are increasing. The largest herd, numbering some 17 000 animals, remains in Wood Buffalo National Park, moving between the grasslands and the boreal forest.

The age of human predation upon the bison has come and gone, and the wolf remains its only natural predator of consequence. Adult bison are simply too large for anything but a pack predator to attack, and even wolves concentrate mainly on the young, the sick, and the old. Historically, the length and intensity of the prairie winter was probably the main limit upon the bison's numbers. Although size is also the best defense against cold weather, calves can only grow so large during their first summer, and their first winter is a test not only for them, but for their mothers, who must find enough forage to sustain themselves while nursing their young. Historically, hundreds of thousands of calves starved to death each winter.

Today, available grazing area limits bison herds. The vast grasslands on which they fed are all but gone, covered by cities and farmland.

LEFT: *The mountain goat sheds its winter wool for a lighter summer coat. The heavy winter coat is necessary because the exposed faces where the goats spend much of their time are cold and windy. When snow forces the goats onto the valley floors to find forage, the creamy white coat hides them against the snow. In summer, when the goats are exposed to direct sunlight but are nearly inaccessible to predators, the coat reflects most of the sun's heat.*

ABOVE: *Musk oxen eat very little for their size and can survive on only a sixth of the forage needed to sustain a domestic cow. One reason their metabolisms are so efficient is that less food energy goes into heating their well-insulated bodies. The guard hairs of their coats are so fine that the Inuit weave them into mosquito nets. A kilogram (2.2 pounds) of the underwool can be spun into a strand 35 kilometres (22 miles) long.*

Domesticated bison may live up to 40 years, although their wild counterparts would rarely survive half that long.

Farther north, the musk ox occupies a similar ecological niche, feeding on willow shrubs, grasses, and sedges. Although considerably smaller than a bison, the musk ox is even better able to withstand the cold and privations of winter, thanks to its thick, shaggy coat. The bulk of the winter coat is wool—possibly the highest quality of any animal's—overlain by a cascade of hair that reaches almost to the ground. It covers most of the animal's legs and tail, which would otherwise lose body heat to the air. In summer, the woollen undercoat is shed.

Musk oxen further conserve their energy throughout the long, Arctic winter by loitering in compact herds of 20 to 30 animals. They congregate on the tops and sides of hills where the snow is shallower, pawing through it with their hooves to find whatever vegetation they can. As with most hoofed animals, grazing is an almost constant job. In winter the caloric returns are so low that it is better to minimize even this low-level activity.

As with the bison, the wolf is their main predator, although desperate grizzly bears will occasionally attack musk oxen—usually the calves. Adults defend their young from wolves by forming a ring with their tails inward and their sharp horns facing outward. Calves shelter in the hub. Lone adults will wade into water to a depth at which wolves would have to swim and so be unable to dodge the musk ox's lethal horns. Musk often evade these predators to live long lives, up to 25 years.

With no place to hide on the open prairie or tundra, musk oxen and bison resorted to intimidation through bulk, but other members of the cow family have retreated to the mountains in an effort to escape their predators. Mountain goats occupy the mountains and isolated river canyons of British Columbia and western Alberta. Bighorn sheep live throughout the western mountains from northern Mexico to the Rockies of southern Alberta.

The climbing skills of the mountain goat are astonishing. Much of their ability is born of their muscular, compact bodies, which are better adapted for single leaps and pulling themselves up near-vertical rock faces than for running. From a distance, the nearly upright posture of a mountain goat scaling a cliff resembles some white ape. Their cloven hooves consist of a hard rim surrounding a soft, almost malleable, pad. With these, they can find a purchase on the tiniest nubs of rock. Their sharp eyes can accurately gauge distances to the next hoofhold.

But as with human climbers, it seems that confidence is what carries many sheep and goats from one ledge to the next or ricocheting down the sides of a rock chimney. The thought of falling doesn't seem to enter their minds. Mountain goats and sheep leap chasms with abandon, and have even been seen doing front hoofstands by walking their rear legs up vertical faces to reverse their direction on narrow ledges.

Of course, mountain goats and sheep do sometimes fall to their deaths—especially the less-experienced kids and lambs. Avalanches and rock slides kill many of these climbers, and predators are only too eager to assist in death by gravity. Golden eagles sometimes knock young—and occasionally even adults—from ledges by striking them with balled talons. They then fly to the valley bottom to feed on their prize. Barring such accidents, mountain goats and sheep often live as long as 15 years.

The reward for the sheep and goats' risky method of predator evasion is superior grazing. Although little vegetation grows on the cliffs themselves, climbing affords sheep and goats access to benches beyond the reach of other grazers. Such pasture may be in small pockets capable

OVERLEAF: *Bison are the largest members of the cow family, and also the largest of all land animals in North America. A bull may stand 2 metres (6 feet) at the shoulder. Most of their weight is concentrated in their massive heads and shoulders, which the bulls use to intimidate predators, and each other in disputes for rutting privileges with bands of females.*

ABOVE: *The woodland caribou is a larger subspecies of the barren-ground caribou, inhabiting the boreal forest and alpine tundra of the Selkirk Mountains from southern British Columbia through Washington and Idaho. Never numerous, this subspecies is now endangered. Many died in collisions with automobiles when a highway was built through their range in the 1960s.*

RIGHT: *Mule deer does lick their fawns dry shortly after birth and move them to a secluded spot, reducing the chance of a predator tracking the scent of their birth fluids to the source. This fawn is about a month old, but the mother continues to lick her offspring, just as a human mother bathes her baby.*

of supporting only a few large animals, so goats are less social than other ungulates and tend to travel in smaller herds. In winter, they descend to the valley bottoms where the snow isn't as deep, vegetation is easier to uncover, and the risk of being swept away by an avalanche is reduced. Bighorn sheep, which are slightly less adept at climbing than goats, descend more readily at any time of year. They congregate in larger herds that afford them some protection from their predators—cougars, wolves, coyotes, bears, and the occasional wolverine.

Dall's sheep are smaller than the bighorn sheep, and their more northerly range sometimes takes them far out into the tundra, away from the cliffs and slopes where they are safest. Because of their more open habitat and greater vulnerability to predators such as wolves, their herding instincts are stronger.

Sheep are far from defenseless. Being rammed by a pair of spiraling horns with the weight of an adult bighorn (up to 160 kilograms, or 350 pounds) behind them is intimidating enough on level ground. On steep terrain, the prospect of jousting with a ram is even more perilous. Sheep often defend their young courageously, sometimes driving off wolves that have taken lambs in their jaws by butting them repeatedly.

Bighorn sheep and musk oxen employ their horns in ritualized skull-to-skull charges. These contests have a dispassionate and controlled air to them, like a human pistol duel. Eerily silent, the males square off at a distance of 14 to 18 metres (15 to 20 yards) and then, as if by agreement, crash head-on with a crack like two curling rocks. Dall's sheep embellish the charge by standing up before falling together.

It's all done in the interest of gaining access to harems. The females spend the year in bands of up to 15 related females and juveniles. Females over three years of age will be receptive to the victors. How they—or even the participants—determine the winner of these skull-ringing duels is unknown. In the case of musk oxen, the cows take a keen interest in the outcome, often lining up to watch. The show is, after all, as much for their benefit as it is for the males'. Head-butting is actually the last resort of evenly matched males. Usually, the contest is ended before it ever begins when a ram or bull demonstrates his superior size simply by displaying his profile to his rival.

By comparison, the bayonet-like horns of the mountain goats are all business, weapons used by both males and females to gore predators and to defend grazing areas from other goats. Females will also protect their young from billies who may try to push them over cliffs. During the rut, the males approach the females on bended knees and are chased away by them immediately after copulation.

Six months later, the nanny retreats to a remote cliff to give birth to a single kid, occasionally twins. She nurses the young goat for eight to ten weeks, but shepherds it about the cliffs until the following year when her next kid is usually born.

It hardly looks like a creature of the mountains, but the peccary of the southwestern United States sometimes forages at high elevations

LEFT: *With her daggerlike horns, this Dall's sheep ewe is quite capable of defending her lamb against predators. The rams have much larger, spiraling horns. These are more useful for their ritualized head-butting displays which take place in the late fall. Like antlers, curled horns are designed to settle dominance disputes between males while avoiding lethal wounds.*

throughout its range, which extends from Arizona to Argentina. Once, they lived as far north as Arkansas. The peccary looks like a small, tusked pig weighing from 14 to 30 kilograms (30 to 65 pounds), covered in bristly, charcoal-colored fur. In fact, it is only distantly related to pigs, having fewer teeth and a diet containing more vegetable matter. The peccary's foot bones are fused, an adaptation for running. Although they have rather short legs for a hoofed animal, they can reach 40 kilometres (25 miles) per hour in a sprint.

Peccaries travel in herds of 6 to 30 animals, browsing on prickly pear cactus, mesquite fruit, sotol, and other plants of the arid southwest. Sometimes, they dig for roots and tubers and only very occasionally supplement their diet with frogs, lizards, or other small animals. In each herd, there are roughly equal numbers of females and males, one of which is a dominant boar. As each of the females of the herd comes into heat, he mates with her, keeping all other males away from the receptive sow for a few days. On those occasions when the dominant boar is challenged by another male, they charge each other, biting and locking jaws. As food is in equal abundance almost year round in the peccary's range, mating can take place in any season.

Most of the time, however, the peccaries co-exist in peace, grooming each other and grunting softly as they forage. They are most active in the cool of the evening and at night during the summer, shifting to a more strongly diurnal lifestyle in cooler weather. They are quite skittish, and normally run from their enemies—cougars, bobcats, and occasionally golden eagles. These predators are not common in their range, and peccaries will fight if they can't escape. Their tusks, which are constantly whetted by the opening and closing of the peccary's mouth, are effective weapons. As a result, if they survive their first year, peccaries live relatively unmolested by predators, for as long as 20 years.

Pronghorns are the only surviving species of a whole family of antelope-like mammals that once roamed North America, but are now extinct. At home on the plains, the pronghorn has little to fear from any predator it sees coming because it is the fastest land animal on the continent; only the cheetah of Africa is faster. Pronghorns have been clocked at 98 kilometres (61 miles) per hour in a sprint and they can cruise for five minutes at three-quarters of this speed.

Like all the hoofed animals, their legs and neck are spring-loaded. As the running pronghorn puts weight on its hoof, a ligament running down

ABOVE: *Pronghorns once roamed the plains of North America in herds that rivaled those of the bison, numbering perhaps 35 million before the arrival of Europeans. Professional hunters slaughtered them for their meat, ranchers shot them as competitors for cattle range, and agriculture removed much of their natural pasture. By the 1920s, there may have been as few as 15 000 prong-horns. Today, their numbers have rebounded to about half a million.*

the back of the lower leg is stretched to its full length. As it pulls the leg back and the hoof leaves the ground, tension in the ligament snaps it back, putting an extra kick into its stride. (You can actually hear this ligament snapping as a caribou runs.) This spring-loading is the reason that the legs of ungulates look so spindly: The bulk of the muscles powering the legs is close to the body, in the rump and shoulders. The muscle mass at the hoof end of the legs is low, reducing the amount of energy needed to swing them back and forth.

The pronghorn has a smaller stomach but larger heart and lungs than other hoofed animals of similar size. Perhaps because of its shorter digestive system, it browses on more nutritious foods than other ungulates, concentrating on shrubs, forbs, and sagebrush. It once shared its range with the bison, which ate mainly grass.

Like the deer, to which they are only distantly related, female prong-horns hide their young for about three weeks after giving birth—usually to twins. The reason for this behavior may be the pronghorn's complete reliance upon speed to escape predators. If the kids were to join the herd before being able to run, predators would simply cut them away from the rest. As with the caribou, concealment probably offers the young better odds of survival. They usually live only four to five years in the wild, although captive animals have survived to three times that age.

In the far north, the caribou shares grazing rights to the tundra with the musk ox. The caribou's hardiness is partly due to its stomach, which hosts special bacteria enabling these animals to digest lichens, the mainstay of their diet. Unlike the musk oxen, caribou range widely in their search for food. In summer, they graze on the grasses, sedges, and lichens of the open tundra. In winter, they move south into the northern reaches of the boreal forest, where they can browse on vegetation above the snow and seek shelter from the cold. The George River caribou herd in northern Quebec and Labrador numbers 600 000 to 800 000 animals, the largest caribou population in the world. Its annual migration takes the herd on a great loop 4000 kilometres (2500 miles) long between its summer and winter ranges.

Caribou must stay on the move to avoid stripping the land bare of vegetation. Although they are the smallest of the northern ungulates at an average weight of about 110 kilograms (240 pounds), a herd requires a

BELOW: *Both male and female caribou have antlers. The annual growth of their racks drains their bodies of phosphorous and calcium, minerals that may be hard to come by on the tundra. Why females should take on this added metabolic burden isn't exactly clear. It may be that on the open tundra the antlers are indispensable for defense against predators. They may also give the females a fighting chance in their disputes over forage with young bucks.*

large foraging area. Migration also gives them relief from their major predator, the wolf. Wolves den at about the same time of year the caribou give birth to their calves. But wolf pups are much less mobile than caribou calves, which are up and walking within hours of their birth. The caribou's summer migration takes them above the tree line and beyond the reach of many wolves at this critical time of year.

The pregnant females of most herd animals give birth en masse over a few weeks in spring. This is thought to swamp predators with calves so that wolves, coyotes, and wolverines can only take a limited number of young before glutting themselves. Often, though, caribou mothers become quite cantankerous and leave the herd to calve. The consistency of the caribou migration attracts wolf packs, which often meet up with a herd like clockwork. Perhaps a calf's chance of surviving is greater away from the herd and its entourage of predators.

The caribou's primary adaptation to this nomadic life is its hooves, which are broader than any other deer's. With them, caribou can walk atop snow and muskeg. In winter, hair growing between the pads keeps them from slipping on ice and hard snow. Unique among the deer family, both males and females grow antlers, although the male's are much larger. They use the antlers not only to defend their young from predators but to sweep away snow while foraging.

Caribou migrations follow consistent routes, fording major rivers. Hundreds of animals sometimes drown in these crossings, but the herd continues. The caribou are to the Inuit what the bison once were to the Plains Indians—a sustaining tide of meat, hide, and bone. In a landscape where wood is almost unknown and where even stone lies buried in snow for most of the year, animals are not just food, but the sole source of material for tools and clothing.

RIGHT: *In keeping with their shy natures, white-tailed deer do most of their browsing in open areas at night. They bolt their food, tearing leaves from branches and swallowing with very little chewing. This minimizes the time spent exposed to predators. At dawn, they seek the security of heavy brush or the skirt of a tree where they can chew their cuds and continue digestion at leisure.*

The migratory, barren-ground caribou are holding their numbers, but the more sedentary woodland caribou populations have been falling since the arrival of Europeans in North America. These animals live in much smaller groups, staying within the boreal forest year round. One theory for this decline is that the northward spread of the moose's range has brought more wolves into contact with the woodland caribou. A caribou's life is not an easy one, and most live only four or five years.

Elk are similar in form to caribou but average twice their weight. This is a reflection of their more southerly range and varied diet. Elk cannot graze exclusively on grasses and herbs, but must also browse on the leaves of woody plants. They even eat bark, particularly that of aspens, when no other food is available. These foods are necessary to provide them with the calcium and phosphate they need to grow their enormous antlers. Elk are highly adaptable, inhabiting deciduous and coniferous forests, swamps, clearcuts, and grasslands in a broad band reaching from just north of the central Mexican border through northern British Columbia and Alberta.

During the rutting season, the bulls proclaim their size and strength with a bugling call that can be heard for miles. The bigger the bull, the louder the bugle. It is, in effect, a boast to females that the caller has

ABOVE: *Bighorn rams settle dominance disputes by charging at each other in head-cracking duels. To withstand such punishment, the skull has two layers reinforced by connecting joists of bone. One noted biologist witnessed two rams jousting continuously for over 25 hours, colliding an average of five times an hour.*

LEFT: *We think of bison as animals of the open plains, but today the world's largest herd lives in Wood Buffalo National Park, on the border of Alberta and the Northwest Territories. These bison spend much of their time grazing in meadows of the boreal forest. A few biologists recognize these "wood bison" as a distinct subspecies, but there seems to be no genetic basis for their belief.*

successfully accumulated a lot of weight and has enough spare energy to brag about it. Strong, silent types may be born into elk populations, but they don't enjoy much mating success.

The goal of all the shouting is to attract a harem of females with which to mate. For their part, the females come shopping for the most suitable father for their calves. The bull's call helps them to gauge his reproductive fitness, and eventually to select a mate. Once they are within sight of the bull, they further assess his strength by the size of his body, especially the rack. Unlike horns, which are permanent and made of dead tissue, antlers are living bone that must be infused with blood and nerves. Bulls grow and shed their antlers afresh each year, and so the rack is a tremendous drain upon their bodies. Like loud bugling, a large, symmetrical rack indicates a fit specimen with resources to spare.

But males within earshot are also listening, assessing the depth and strength of their competitor's call. If they deem themselves capable of meeting the challenge, they too will approach the bugler, sizing him up in much the same way the females do. Through friendlier sparring matches outside the rutting season, bulls are aware of their own size and fitness. The onus is on the challenger to prove his superiority, and he can only do this by attacking. If he can drive off his rival, he inherits the harem.

And so the bulls fight, each using his own rack to engage the other's. Antlers are not primarily designed to inflict wounds, as might first be surmised from looking at all those sharp points, but to grapple. The many tines provide forking points for the racks to seat firmly against one another. Once joined, it becomes a wrestling match, designed to test the combatants' strength.

All of the shouting and fighting comes with a high price tag: It attracts predators. Such a noisy reproductive strategy is only practical for animals

Left: *"Elk" is actually the British word for moose. Early settlers mistook the North American elk for a moose and so applied the misnomer. Its other common name, wapiti, comes from a Shawnee word meaning "pale deer" and is a more apt term.*

ABOVE: *Most animals living near the poles are larger than their equatorial cousins, but above a certain latitude there isn't enough food to sustain large land animals. Peary's caribou, which lives as far north as Ellesmere Island and the northern tip of Greenland, is smaller than the woodland or barren-ground caribou.*

RIGHT: *The horns of the pronghorn share the characteristics of both horns and antlers. The inner core is composed of bone, but a sheath of keratin, the same material that hair and hooves are made of, grows over the core and is shed annually. This shot also gives an excellent view of the pronghorn's large, protruding eyes, which give it an extremely wide field of view. The pronghorn can spot movement up to 6.5 kilometres (4 miles) away.*

that are either large enough to intimidate or small enough to hide. Elk are large enough to intimidate, and are the only deer that bugle during rutting. They're so aggressive at this time that most predators think twice about attacking. The antlers which can safely lock with another rack become much more dangerous weapons when used against a predator that is not so equipped. Thus, the bugling of the bulls broadcasts a third message: If you aren't an elk, stay away.

One of the bulls eventually tires and concedes defeat by retiring. Elk do gore one another, and when this happens, combat is broken off when the wounded party signals his defeat by turning aside or even sitting down. Occasionally, the combatants' antlers become locked. Unable to disengage, such animals either become the victims of predators or starve to death.

Hiding in the woodlands of North America are the remaining members of the deer family: the mule deer, white-tailed deer, and moose. They remain out of sight of their predators by foraging in open areas at night, and running for cover if disturbed. They spend much of the day bedded down in a thicket or under a tree, chewing their cuds. The forest and its margins are an ideal place to practice this hide-and-seek strategy. Mule and white-tailed deer do not have the stamina of the pronghorn or the

bison; they need to run only far enough to be lost among the trees. White-tail deer establish a network of trails within their relatively small ranges and become highly familiar with them. They know just which way to run when a predator comes after them, and this strategy is especially effective in deep snow. By packing down a trail, a deer that would normally be floundering up to its chest can escape.

Mule deer practice a different escape strategy. They too are intimately familiar with their home ranges but when pursued by predators, they deliberately bound over logs, fences, boulders, or other obstacles. Exceptional jumpers, they quickly leave their pursuers behind. Black-tailed deer, a subspecies of the mule deer that lives on the west coast, can move quietly through the densest brush, and rely most heavily on concealment to elude their predators. If flushed, they also make their escape path an obstacle course by heading straight for boulders and logs and leaping over them. Their pursuers may be wolves, coyotes, cougars, lynxes, or bobcats. Deer are also the hoofed animals most frequently sought by human hunters. Every year, millions of them are shot and almost as many are accidentally killed in collisions with automobiles. They're a popular prey item because of their abundance and because they are of more manageable size for a predator than elk or moose. Still, a big mule buck may weigh as much as 200 kilograms (450 pounds) and a white-tailed buck 140 kilograms (300 pounds). The does of both species are considerably smaller, and without antlers are even more desirable to non-human predators.

The deer of the forest are generally silent and solitary. White-tailed deer have prolonged contact only as mother and fawn. Mothers that have given birth to more than one fawn will even separate the siblings, sometimes by as much as 90 metres (300 feet). Separating twins is a strategy

LEFT: *For most of the year, the moose is the shyest of all deer. But the bull moose becomes a different animal during the fall rut and many hunters and naturalists consider it the most dangerous animal in the forest. A rutting moose charges through the brush in search of rival bulls and has been known to attack people, automobiles, and even a freight train on one occasion.*

for preventing a predator from killing both her offspring. A coyote or a cougar having eaten one fawn is unlikely to seek another immediately. Newborn white-tails are able to stand within hours of birth, but for their first few days they lie in quiet concealment until their mothers return from browsing to suckle them. Until they can run, she keeps away from them between feedings, as her scent and even the faint sounds of her browsing would risk alerting predators.

Mule deer frequent more open areas than the white-tailed deer and are more gregarious, seeking the protection of the herd when grazing. They have larger ears for picking up distant sounds. Like other deer, they often follow the snow melt in mountainous areas, browsing on tender vegetation in spring. While white-tails are found in wooded areas right across North America, the mule deer is mainly a western animal. They have lived as long as 25 years in captivity, but seldom survive more than 10 in the wild.

Moose are second in size only to the bison and may weigh as much as 800 kilograms (1750 pounds.) Despite their size, they are as shy as any other deer and seldom stray far from a pond or lake. They live almost anywhere north of the Great Lakes where there are trees, except for coastal British Columbia. In summer, moose do much of their browsing on aquatic plants and are good divers, sometimes foraging completely submerged. They can outswim wolves, their chief predators, and like the musk ox will retreat to water if harassed. They will even head for the shallows of a creek when pursued.

While most deer attempt to outrun predators, observations of moose attacked by wolves indicate that those standing their ground are more likely to survive. If it can't enter water, the moose tries to back against thick brush or some other obstacle and attempts to kick or stomp the wolves with its front hooves. In one eyewitness account, seven wolves all had a grip on a moose at one time, one on the moose's nose. The moose managed to shake all of them off by entering water, and eventually escaped.

The moose's long legs are as useful for wading through deep snow as they are water. Still, like all deer, it is most vulnerable to both predators and starvation in winter. If winter is such a hardship for the hoofed animals, why don't they sleep through the cold, hibernating as so many other animals do? Unfortunately for ungulates, size and long legs—the very attributes that deliver them from their many predators—leave them ill equipped to hibernate. Animals that hibernate must protect themselves from exposure by burrowing underground or entering caves or tree cavities where they can shelter from the cold. Long legs are a hindrance when trying to squeeze into a small space. Also, hooves make poor shovels; excavating a den large enough for a moose or a bison just isn't practical.

Some large animals do hibernate—grizzlies can scoop out a den big enough to winter in. But their powerful forelimbs, used for tearing apart stumps and unearthing rodents from their dens, are naturally adapted to digging. A body form that is good at both running and digging is an unlikely evolutionary accident. The bear is probably the only animal bigger than a dog that does both well.

The growth of the cloven-hoofed animals into the largest and fastest of terrestrial animals was an evolutionary success. For millennia, they darkened the plains with their herds. But the tales of evolution have their ironies. When human beings reached North America, the bodies that had served the hoofed animals so well were turned in service of their new predators: The bones and sinews that had given the ungulates their stride became the hunters' tools, the meat of their muscles sustenance, and their hides shelter from the cold and wind.

Short Legs and Tempers

The weasel or mustelid family is a large group of carnivores that have exploited a variety of habitats. Martens climb trees. Otters are almost as comfortable in the water as seals. Wolverines walk over tundra and weasels snake through the burrows of their prey. In adapting to these habitats, evolution has stretched the mustelid body like taffy or kneaded it into a squat loaf. But looking at those bodies, the impression is that their designer insisted on one, ironclad specification: Keep the legs short.

With the exception of the wolverine, the North American mustelids scuttle along with rapid steps of their little legs. They share other qualities: Almost all are fierce and solitary predators; they are intelligent, curious, and often playful; and all except the sea otter have anal glands that produce a strong, musky scent.

The weasels, of course, are the definitive members of the family. Their bodies are like pipe cleaners, allowing the weasels and their close relatives, the mink and ferrets, to track rodents, shrews, and rabbits down their holes. They try to corner their prey in a blind tunnel before killing them with a bite to the neck. Weasels also eat birds, some insects, snakes, and occasionally berries.

ABOVE: *Many animals with chemical defenses bear distinctive markings that predators can easily recognize and avoid. Most of a striped skunk's potential predators are mammals having monochromatic vision, hence its contrasting black and white pattern. This pattern is also more visible at night, when skunks are most active.*

LEFT: *Martens hunt primarily for voles and other ground-dwelling rodents, but in spring when squirrel and bird nestlings are abundant, the marten's semi-retractile claws allow it to take to the trees and exploit arboreal food sources. Like all members of the weasel family, it has a voracious appetite and must feed often.*

Weasels and other slender-bodied mustelids must eat frequently, and only animal protein is sufficient to maintain their high metabolisms. Their long, thin bodies do not retain heat as well as the squatter bodies of badgers or skunks. And yet, these carnivores often live in the coldest parts of the continent. Short-tailed weasels live throughout Canada right to the tip of Ellesmere Island, in Alaska, around the Great Lakes, and in the western mountains of the United States.

The least weasel is the smallest of the true carnivores, which compounds the problem of maintaining body temperature. To compensate, it has a metabolic rate equivalent to that of some shrews, and may have to eat half its weight in food every day. Weasels usually move into the burrow of an animal they have killed. To further combat heat loss, they may line their tunnels with grass and sometimes, in a grisly touch of interior decorating, the fur of their original owners.

Weasels rely on speed to overcome larger animals. They often wrap their tubular bodies around rabbits five to ten times their weight more in the manner of a snake than a mammalian predator. As the animal struggles, they work their way to the back of the neck where they deliver a killing bite. During the summer, weasels will often cache the bodies of rodents in much the same way that squirrels and chipmunks save nuts. They will use this stockpile to supplement their winter kills. Over a hundred rats and mice have been found in a single cache.

The coats of weasels are brown in summer, but turn white in winter. Their white coats camouflage them as they tunnel through snow in search of prey. But the tip of the weasel's tail remains black year round. Studies of the long-tailed weasel have shown that this black tip distracts birds of prey into attacking the tail, frequently allowing the weasel to escape. One theory is that the birds mistake the tail for the weasel's head, and so

RIGHT: *The sea otter relies on air trapped in its fur, made water repellent by oil smeared from its anal glands, for buoyancy. Oil spills are devastating to the sea otter because, in addition to the usual toxic effects, petroleum destroys the ability of the otter's pelt to repel water. Soaked to the skin, the animals become susceptible to hypothermia and drowning.*

are taken by surprise when the weasel bolts in the wrong direction.

The black-footed ferret is a weasel that specialized in preying on prairie dogs. It bred and hunted within the confines of their once huge colonies. Unfortunately, the extensive poisoning campaigns directed toward the prairie dog were even more effective in eradicating its predators. Biologists captured the last 18 wild black-footed ferrets in 1987 in Wyoming. They have managed to breed the ferrets in captivity, but their efforts to re-introduce them to the wild prairie have been less successful. After releasing 23 in a prairie dog colony near Fort Belknap, Montana in 1997, only 3 could be found the following spring. The next year, survival over the winter improved to 11 out of 25 released. Some may have dispersed to other locations, but biologists are still mystified as to the fate of the lost ferrets. For now, the black-footed ferret remains the rarest mammal in North America, and the nine-digit identification number recorded on a microchip implanted in each of their necks seems a desperate act of optimism.

The marten is a weasel that climbs the trees of North America's coniferous forests, scampering from branch to branch on paws with retractile claws. Although it hunts mainly on the ground, it is fast enough to catch squirrels in the forest canopy. Like all the true weasels, it does not

ABOVE: *During the winter, the coat of the short-tailed weasel, or ermine, is white, providing camouflage against the snow as it searches for rodents, shrews, and other small mammals. In summer, its coat turns dark brown except for the white belly and neck. The British refer to the ermine in summer as a stoat, but the term is not common in North America.*

LEFT: *The distinctive, elliptical entrance of an American badger's den is the natural result of the badger's flattened body. With their strong legs and heavy claws, badgers are among the best diggers of all mammals, and their burrows are much in demand as dens from other, less-proficient burrowers. Few are foolhardy enough to try to move in while the badger is still in residence. This rattlesnake would seem to be an exception.*

hibernate, hunting right through the winter. The marten is highly carnivorous, but will supplement its diet with berries in the summer.

The fisher is much larger than the marten—about the size of a fox—and has been known to catch and eat its smaller cousin. It does not fish. It has a number of adaptations for climbing trees, including semi-retractile claws and a swiveling ankle joint, which enable it to descend trees headfirst. This is useful when attacking porcupines, whose natural tendency is to escape predators by climbing trees. Because the fisher can descend the trunk headfirst, it can scamper up a tree, then turn and attack the porcupine's unprotected face. The fisher does most of its hunting on the ground, however, patrolling a range that is usually about 30 kilometres (19 miles) in diameter over a week or so, searching for rabbits, rodents, birds, fish, snakes, and insects. They have been trapped for their fur to the point where they are rare animals, greatly benefiting the porcupine population of North America. Fishers are still found in a broad band of coniferous forest crossing Canada and the northeastern states.

Mink are weasels that live close to water, hunting underwater for fish, frogs, crayfish, worms, and aquatic birds. They also eat rodents, ducks, songbirds, shrews, snakes, and some insects. Their feet are partially webbed and their thick fur, as a result of the oils secreted from the mustelid glands to keep it waterproof, is particularly lustrous. The most versatile hunters of all the mustelids, mink also hunt on the ground and even in trees from time to time. There were once two species in North America, but the sea mink, common along the Atlantic coast, was exterminated by fur traders just before the turn of the century. The mink's glossy fur is still popular with furriers, who require the pelts of about 100 of these small animals (an adult mink weighs about 1.6 kilograms, or 3.5 pounds) to make a full-length coat.

Right: *Despised by trappers because of the damage it may do to their catches, the wolverine itself is sought for its pelt. Its coarse, long fur is valued as trimming for the hoods of parkas because of its natural ability to shed the frost that accumulates from the wearer's breath in cold weather.*

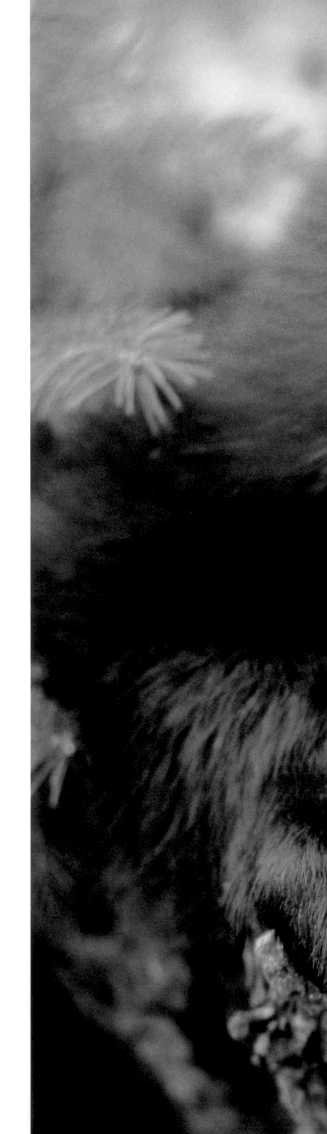

ABOVE: *The river otter is no friend of the beaver. Not only will otters usurp beaver lodges by attacking and killing the original builders, they have been known to dig holes in their dams to drain them, concentrating fish and other prey in the depleted reservoir.*

As adept as the mink is at underwater hunting, it's a landlubber compared to the otter, which is more thoroughly adapted to aquatic life than any mammal but the seals, dolphins, and whales. Both sea and river otters are such good swimmers that they can actually chase down slower fish.

The river otter's body bears the marks of its aquatic life: Its feet are webbed and it has proportionately larger lungs than a mink. It keeps its dives short, usually under 15 seconds, and so the larger lungs may be more for improved buoyancy than dive endurance. For protection from the cold, river and sea otters rely on the thickest and most prized of animal furs. As with the fur seals, outer guard hairs trap air within the underfur, preventing water from reaching its skin. But an otter's fur is nearly twice as dense as a fur seal's. River otters are much larger than most other mustelids, up to 14 kilograms (30 pounds), which further helps them to retain body heat.

The river otter lives near swift mountain streams, rivers, marshes, ponds, or estuaries and forages along their margins, eating almost any animal smaller than itself. Its elongate trunk and short legs give the otter an awkward gait, yet it will sometimes travel for hours overland between bodies of water or to find food.

Not strong diggers, river otters usually take over the vacant lodge or den of a beaver to raise their young. A pair of otters was once found sharing a beaver lodge in New York State with several beavers. As otters have been known to drive beavers from their lodges and eat their young, biologists were at a loss to explain these strange bedfellows.

River otters give birth to one to six pups, blind and helpless for the first three weeks of their lives. At somewhere between six and nine weeks, they take their first swim in the company of their mothers. Otters in general, but especially the young, are famous for their exuberance at play. Siblings romp with each other for hours, slipping in and out of the water and using muddy riverbanks as slides, over and over again. Most young mustelids are highly playful, echoing a strong correlation between the ferocity of adults and their playfulness as juveniles that reverberates through the whole class of mammals.

The river otter's bigger marine cousin, the sea otter, rarely leaves the ocean. The female even gives birth to her young—usually a single pup— in the water, bending double to assist in the delivery by pulling on the

BELOW: *The mink is truly at home in the water, able to dive to depths of at least 5 metres (16 feet). It often dens in the burrows of muskrats, its preferred prey, but occasionally digs its own. Mink live along rivers, ponds, creeks, lakes, and marshes throughout Canada and the United States except for the southwestern deserts and the extreme north.*

emerging head with her teeth and front paws. She then rolls over onto her back and licks the pup dry of amniotic fluid. For the first two weeks of its life, until its eyes open, the mother carries her pup on her chest or back. Shortly thereafter, the young otter begins to dive and swim on its own. After weaning her pup, she catches most of its food—abalone, crabs, sea urchins, and a variety of shellfish living in the kelp forests along the near inshore of the Pacific coast—for up to a year.

Sea otters are the only mammals aside from primates to use tools. Some keep a stone in a pocketlike fold of skin under the armpit and use it to help dislodge shellfish. It may take an otter repeated dives and pounding with the stone to remove a really stubborn shellfish such as the abalone. Once on the surface, the otter rolls onto its back, sets the shellfish on its chest, takes the stone in both paws, and uses it to hammer at the shell until it cracks. Otters also tie a strand of kelp around themselves to keep from drifting out to sea while napping. At least one male was known to chew through discarded beer cans to reach the young octopuses hiding inside.

Once, the sea otter's range included the whole North Pacific rim, and their population may have been as high as 300 000. Fur traders reduced their numbers to fewer than 2000 animals at the beginning of this century—not surprising when a single pelt could sell for $1000. Their habit of remaining close to shore and their short dive times (less than two minutes) compared to other marine mammals made them easy prey. In 1911, in a rare and early example of environmental cooperation, international legislation put a halt to the hunt. Today, sea otters occupy about half of their historical range and are most numerous in the Gulf of Alaska and along the Aleutian Islands. Otters taken from stock in northern British Columbia and a remnant group in California have been re-introduced to parts of the California coast and appear to be doing well.

LEFT: *The skunk is the only mustelid that uses its scent glands for defense. Most members of the weasel family use their glands to mark their territories. The marten, shown here, drags glands on its underside over branches and logs to establish what are called scent posts. These serve as a no-trespassing sign to other martens.*

RIGHT: *With its webbed feet and rudderlike tail, the river otter is an agile pursuit predator, capable of chasing down slower fish, but occasionally catching fast swimmers such as trout. The otter can hold its breath for several minutes and has been known to dig itself into mud or sand at the bottom of a body of water to ambush fish. Unlike the sea otter, it usually eats its catch ashore.*

If any mustelid could escape the ravages of the fur trade, one would think it would be the skunk, but even they have been trapped for their pelts. For obvious reasons, skunks have very few natural predators. The source of their notorious odor is a modified musk gland about the size of a cherry that opens through a duct exiting the anus. When threatened, the skunk can voluntarily squirt the contents of the gland about 5 metres (15 feet), with good aim at about half that distance. The substance is quite a strong poison but not actually dangerous unless ingested.

The smell alone is enough to deter all but the most desperate predators. Birds of prey, which have poorly developed senses of smell, and hungry fishers, cougars, or bobcats will sometimes attack skunks. For most animals, though, seeing the skunk's distinctive color pattern is all that's necessary. If that doesn't work, the skunk has an escalating dance it begins as soon as trouble approaches. First it walks toward the intruder with tail erect. It clicks its teeth and hisses. If that isn't sufficient, it will stamp its front feet. The spotted skunk finally does a handstand with its tail erect.

Spraying a predator is a last resort for two reasons: Replacing toxic or noxious chemicals spent in an attack is metabolically costly and it leaves nowhere for the skunk to go. Once a predator is sprayed, it has little to lose and everything to gain by finishing what it started—if it can still function. For the most part, the skunk seems supremely confident of its weapon. Young skunks engaged in play will sometimes chase each other right up to people, completely heedless of any danger.

Skunks have keen noses and yet are somehow able to ignore their own odor. Their noses lead them to insects, worms, berries, birds' eggs, carrion, rodents, buds, and grasses, mainly at night. As most of these foods are scarce in cold weather, skunks dig underground dens where they pass the worst of the winter in a torpor, but not a true hibernation.

Living on the dry plains of the United States and the southern prairie provinces, the American badger is an even better digger than the skunk. While the weasels have evolved their sinuous bodies, fitting them for underground pursuit, the badger prefers to dig out its prey—mainly ground squirrels. The badger has an oddly flattened, spade-shaped body ideally suited to the task, and special membranes over its eyes to protect them from dust and dirt. It also has some tricks to prevent its prey from escaping. The badger will sometimes search out and plug an animal's back door before commencing excavation.

This would seem, at first, an inefficient hunting strategy. Digging is a

ABOVE: *As is true of many members of the weasel family, the fisher employs delayed implantation. Although it mates in early spring, for the first 10 to 11 months the fertilized egg is not implanted in the wall of the uterus and so does not grow. Birth occurs about a year after fertilization. The delay may give the fisher's body a time in which to "decide" to spontaneously abort the fetus, depending on how difficult the winter has been.*

great deal more work than running down prey. But the badger's tunneling skills pay it other dividends. While weasels and ferrets must usurp the burrows of rodents for their homes, the badger can construct its own extensive system of tunnels. As well, the strong, compact limbs and claws that make the badger one of the animal kingdom's best diggers also make it a formidable fighter.

The grizzly bear is about the only predator able to dig a badger out of its elaborate burrow, and one of the few willing to battle the maelstrom of claws and teeth that is likely to be its reward. Golden eagles and coyotes will sometimes attack badgers in the open, but coyotes are more likely to turn the badger's talents to their own advantage. They have been known to follow foraging badgers and pounce on anything that bolts from a rear entrance once the badger begins digging.

The wolverine is the oddball of the mustelid family. It lives mainly on the tundra and in the boreal forest and looks like a small, shaggy bear. The male weighs only 14 to 27 kilograms (31 to 60 pounds), but it has longer legs and larger feet than any of its relatives. As if taking the blessing of longer legs to heart, wolverines are excellent walkers and cover enormous home ranges—as large as 700 square kilometres (270 square miles). Wolverines have walked 32 kilometres (20 miles) in a night of scavenging. Carrion is probably the wolverine's main food, but it has earned an unequaled reputation for ferocity by attacking and killing caribou. Because of its large hairy feet, the wolverine can run on deep snow when its prey is floundering, which is the only way it can bring down an animal as large and fast as a caribou.

The wolverine's ferocious reputation has probably been exaggerated somewhat by the trappers whose lines it raids. With its appetite for carrion and the ability to walk for miles, a wolverine may follow a trapline from one end to the other, feeding on the bodies of the animals and spoiling their pelts in the process. For this, it has been called the devil bear, and the Inuit have named it *Kee-wa-har-kess*—the Evil One. Even its scientific name, *Gulo gulo*, means glutton. Could they speak, the many mustelid species that have been driven to—and sometimes over—the brink of extinction at the hands of the fur trade would no doubt have chosen different names for their champion.

BELOW: *In spring, mink are born naked and blind in a nest lined with the fur of rodents and rabbits. Litters consist of 1 to 10 kittens which remain together, honing their hunting skills, until the fall when they disperse to find their own territories.*

Nasty, Brutish and Short

For a shrew, eating is only slightly less urgent than breathing. From the moment it finishes devouring a cricket or an earthworm—animals hardly smaller than itself—it has only a few hours in which to find another before it starves to death.

The voracious appetites of the shrews and moles are driven by their metabolisms, which consume calories with the urgency of a burning fuse. The heart of an active shrew vibrates at over 1200 beats per minute. They need such rapid metabolisms to warm their tiny bodies. Shrews are the smallest of all mammals. The smallest in North America, the pygmy shrew, weighs scarcely as much as a dime. Being so tiny is not a problem for a fish or an insect, animals whose bodies are roughly the same temperature as their surroundings. But mammals must maintain a constant body temperature—often considerably higher than the air or water around them. The smaller the body, the faster it loses heat to its surroundings, and so the need to constantly stoke the metabolic furnace. Although some shrews supplement their diets with seeds and fruits, most must satisfy their appetites with insects, spiders, earthworms, and other invertebrates which they find in the upper layers of soil or the leaf

ABOVE: *The plethora of shrew species is something of a mystery. At first glance, they all appear to be mouse-sized animals that hunt for insects, worms, and other invertebrates. But shrews actually come in a variety of sizes. The pygmy shrew may be only 78 millimetres (3 inches) long (including tail), whereas the water shrew may be up to 158 millimetres (6.3 inches). They hunt a corresponding range of prey sizes.*

LEFT: *The velvety fur of shrews and moles is unique. Each hair is composed of alternating flat and cylindrical segments. The flat segments act as hinges, allowing the hair to bend in any direction. This adaptation to their underground life allows them to back through soil as easily as they move forward.*

litter covering it. Only animal protein provides enough calories to sustain them.

Winter is a difficult time for shrews, and many of them starve. Their bodies are too small to store enough fat to sustain them through hibernation. Shrews of the northern and temperate regions actually respond to the cold with a kind of anti-hibernation: They turn *up* their metabolisms to generate even more body heat, and intensify their search for food. They continue to eat insect larvae and pupae buried in the soil and adjust their diets to include more voles and mice—rodents that also remain active during winter. The short-tailed shrew, one of the best diggers, excavates tunnels in the snow just above the ground. Other shrews use the tunnels of their prey.

Such a precarious life strategy might suggest that shrews are rare or exotic animals, but wherever soils support large numbers of invertebrates, shrews are plentiful. One hectare (about 2 1/2 acres) of forest floor with its productive leaf litter may support 200 short-tailed shrews in summer. Although this is not many compared to the numbers of voles or other small rodents in the same area, shrews may easily make up 25 or 35 percent of the small mammal population.

A shrew's eyes are tiny, even considering its body size, and its vision is poor. As with many animals that see poorly, their hearing is excellent. The wandering, short-tailed, and water shrews have some limited ability to echolocate. Their sonar is effective over only a very short range—probably a maximum of about 65 centimetres (2 feet). In one experiment, shrews were trapped on a tiny platform in complete darkness. The only way down was by jumping to a second, lower platform that gave them access to food. The shrews were found to utter sounds in the 50 to 60 kilohertz range (well beyond human hearing), at first randomly, and

RIGHT: *The shrew mole, as its name implies, shares characteristics of both shrews and moles. It is smaller than most moles but large for a shrew. It also spends far more of its time above ground than other moles. It has other physical traits that aid it in life on the surface, including forepaws it can place flat on the ground and a longer tail which allows it to rear on its hind legs without falling over.*

ABOVE: *Although it spends much of its life underground, the only word to describe the star-nosed mole is unearthly. What appear to be the eyes in this photo are in fact the mole's nostrils. The odd, pink tentacles surrounding them are covered in Eimer's organs, tactile organs that some biologists believe may also be sensitive to electric currents. The star-nosed mole swims as well as it digs, and can be seen moving under the ice covering ponds in the dead of winter.*

then in the direction of the lower platform as they focused in on it. Although their sonar cannot match the acuity of a bat's or a dolphin's and is much less intense, shrews appear to be able to tell whether open space or some kind of object lies in front of them. Such an ability would seem useful in a shrew's world, which is far more cluttered than the spaces in which bats and dolphins hunt.

Of all its senses, the shrew is probably most dependent upon touch. It probes through the leaf litter and upper layers of soil, feeling for movement with the vibrissae (whiskers) on its flexible and sensitive nose. Most of what a shrew eats it probably finds by bumping into it. They're burning so many calories in maintaining their body temperatures that to be in a perpetual search pattern takes hardly any more energy.

Fumbling its way to its next meal would be a disastrous strategy for a cougar or a rattlesnake, but it works for shrews because of their tiny size and the abundance of insects and worms living in moist soils. They can make a meal out of a nematode or an ant—organisms too small for even a mouse to bother with.

Their constant movement gives shrews a frantic, obsessive quality. They often look as if they're searching for lost keys. Shrews sometimes plunge heedless through streamside pools, running along the bottom as if totally unaware of the water around them, and emerge on the other side.

The water shrew, relatively large with a body length of 13 to 16 centimetres (5 or 6 inches), hunts underwater. Its fur traps air so that it swims along in a silvery bubble. It must paddle frantically to remain submerged in its search for small fish, insect larvae, and tadpoles. As its fur is not highly water repellent, it keeps its dives short and immediately dries itself afterwards to reduce heat loss. At times, it may actually scramble across the surface of puddles supported by surface tension and the hairs on its feet.

A few shrews kill vertebrates much larger than themselves, increasing the variety of prey available to them. A meadow vole might be twice the weight of a short-tailed shrew, but the shrew has an advantage: At the base of its incisors is a gland that secretes a neurotoxin. A groove in the incisors channels the poison to the tip of the teeth. The neurotoxin is more disabling than lethal, and it may take 20 minutes until a larger animal such as a mouse or a frog weakens to the point where the shrew can finish the job, usually with a bite to the back of the neck.

Short-tailed shrews cache crickets, earthworms, or insect larvae that they have paralyzed for later consumption. They may place them in a vacant rodent's nest or bury them in the soil, and often mark the cache with urine or feces. This may make the food less palatable for other animals, or help the shrew to relocate it. Not all shrews are known to cache food, but for an animal whose life depends on eating frequently, having a larder makes sense.

Like everything else in a shrew's life, courtship—if you could call the male chasing the female around with his nose over her rump courtship— is done in a hurry. After a few minutes, a female in estrus will relent to the male's efforts to mount her. He grips the hair on her head with his teeth and copulation lasts from a few seconds to half a minute. The male's ejaculate is often followed by a waxy plug to ensure that he is the last to successfully mate with her for this pregnancy. As with wolves and dogs, the couple may be locked together for some time afterward, the male being dragged around by his partner.

It's difficult to know just how long a shrew's gestation period is because, as with many mammals, the female can delay implantation of the egg. Pregnancy lasts from 13 to 25 days, depending on species. Curiously, the larger shrews seem to have the shorter gestations.

The mother builds a dome-shaped nest of grasses, leaves, and moss in

ABOVE: *This eastern mole's natural response to being placed on soil is to dig into it as quickly as possible. It shovels dirt underneath and to either side of its body with alternating strokes of its paddle-like foreclaws. In loose soil, it can disappear completely within five or six seconds. Moles do not always move through soil by digging, but excavate a network of permanent tunnels.*

RIGHT: *Townsend's mole is our largest mole, reaching a length of 23 centimetres (9 inches). It inhabits a strip of coastline from southern British Columbia to Mexico. Few predators can find moles to catch, but weasels may pursue them through their tunnels.*

an underground burrow or a tussock of grass. There, she gives birth to a litter of anywhere from 4 to 11 blind and naked young, with 5 or 6 the most common number. The young of the pygmy shrew may weigh only a quarter of a gram (less than a hundredth of an ounce) at birth.

Competition is stiff for a shrew from the very start of its short life. Although there are enough teats for each to have its own in all but the largest litters, some of the babies begin to grow faster, and some decline and die. The survivors grow rapidly, and in two weeks weigh from 5 to 7 grams (.18 to .25 ounces)—a 20-fold gain in their birth weight. Their eyes open at about this time. They spend much of their time climbing over and burrowing under one another, each trying to get to the center of the shrew pile.

In just 18 days, they're nearly as large as their mother, and suckling them is, quite literally, a draining task. She must eat 125 percent of her body weight each day in order to feed them and herself. By the time they reach three weeks of age, she is actively discouraging their efforts to nurse with churlish calls and threats.

Weaning is a critical time in a shrew's life. It must begin foraging on its own, but from the moment it is cut off from its mother's milk, the clock

is ticking. In a few hours it must learn to feed itself or die. Captive shrews have been seen accompanying their mothers on an initial foraging before soloing, but no one knows if this is the usual procedure.

Young shrews in particular are vulnerable to predators, but they are not without their defenses. Most have glands on their flanks which exude a noxious, foul-tasting musk. Consequently, most other mammals learn to leave shrews alone. Birds of prey, and owls in particular, seem undeterred by the taste and feed heavily on shrews. Snakes will also eat them. Even if they manage to avoid these predators, a shrew's life is brief. Most die shortly after their second summer, living between 12 and 18 months.

The moles have dealt with two of the insectivores' greatest vulnerabilities—heat loss and predation—by going underground. Plenty of mammals live in burrows, but none has adapted so thoroughly to a subterranean existence as the moles. Their lozenge-shaped bodies have been formed for digging in and moving through soil. They're such strong diggers that a mole set on even hard ground can bury itself completely within five or six seconds.

In North American moles, the most radical adaptation has been the forelegs, which hardly function as feet at all. They're more like clawed paddles, splayed to the sides of the body. Moles dig with an action a little like a swimmer's breaststroke.

This movement is made possible by the unusual orientation of the mole's humerus: Instead of hanging down from the shoulder, as does our upper arm bone, the resting position of the mole's humerus is pointing up and out from the shoulder, like a shipped oar. The digging force is provided by muscles that rotate the humerus about its long axis.

The mole's pelt is made up of fine, dense hairs that are hinged near the base so that they may be brushed in any direction with equal ease. This allows the mole to back through loose soil as easily as it moves forward. The eyes of burrowing animals are often subject to infection from particles of soil, but the mole's eyes are either tiny or completely grown over with skin. They're of little use except on those occasions when moles are forced to leave their burrows—usually when a flood fills them with water. Most moles can probably distinguish light and dark, but with the exception of the star-nosed mole, which spends much more of its time foraging above ground, they are probably unable to make out distinct images.

Outside of animated cartoons, the talpid mole family, which includes all North American species, does not move from place to place by swimming through soil. However, some of the golden moles of Asia push through loose sand without ever creating an open tunnel. Even with a mole's extreme adaptations, digging is still hard work compared to crawling.

Moles dig two kinds of tunnels: One runs just beneath the surface, and often leaves a ridge of raised, loose earth. These tunnels collapse behind the mole as it digs and are used mainly for feeding when the weather is mild and the soil is soft. The other kind of tunnel is permanent and runs deeper, as far down as 1.5 metres (5 feet). These tunnels supply the mole with food when cold or dry weather drives the invertebrates living in the soil deeper underground. Moles also retreat to these tunnels for warmth, to give birth, rear their young, and escape predators.

It's now believed that the bulk of a mole's food is collected by patrolling the permanent network, feeding on earthworms and other invertebrates that have blundered into the tunnels. Moles supplement their animal diet by feeding on roots and tubers, and also use dried leaves in their nests. So it turns out there is a kernel of truth to that other cartoon standard of garden plants being yanked underground by the varmint mole.

In addition to protection from its many predators, life underground

offers another enormous advantage to a small animal: a more equable environment. A tunnel is cooler in summer and warmer in winter than the air above ground. As a consequence, the mole's metabolism, while high, does not have to operate at the furnacelike levels of the shrews to keep it warm. Moles are also larger than shrews, which helps them to retain more heat.

Still, a mole is as voracious as any other small mammal of its size. To stay healthy, it must eat about half its body weight in insects and other invertebrates every day. An adult Townsend's mole, the largest species in North America, weighs about 125 grams (4.5 ounces). It would take about 22 earthworms to equal half its weight.

The moles' dependence on tunnels to feed themselves probably accounts for their more southerly ranges. For most moles, the soils of the north and the worms living in them are frozen for too many months of the year. The exception is the star-nosed mole, which spends much more of its time in the water and above ground. It can be found in the eastern half of Canada as far north as southern Labrador.

Moles are solitary creatures for most of their lives, living within territories defined by the extent of their tunnel networks. If an adjacent mole territory is owned by a female, there may be some territorial overlap at the margins, where each mole's tunnels entwines with its neighbor's. Females may even share some tunnels, though seldom at the same time. Male territories are mutually exclusive.

Once all the territories in an area have been established, moles coexist peacefully. Except as infants or when seeking a mate, they take pains to avoid each other. Although moles have no external ears, their hearing is excellent and they probably keep from breaking in on each other's tunnels by listening for the sound of their neighbor's digging. A mole's

Left: *With their tiny eyes and poor eyesight, shrews have almost no ability to detect predators at a distance, which is one of the reasons they avoid open areas. This behavior offers shrews some protection from one of their main enemies, owls. Other predators avoid shrews because of glands along their sides that exude a foul-tasting musk.*

territory may cover from 300 square metres (3225 square feet) where soil invertebrates are plentiful, to 10 times that size where food is sparse. Within that area, moles may dig extensive tunnel networks. One eastern mole dug a single tunnel that ran almost one kilometre (3200 feet) along a fence.

If circumstances should force them together, moles almost always fight. In the wild, these battles usually occur when young moles leaving their mothers' nests make the mistake of trying to dig a tunnel within the boundaries of an existing tunnel network.

In spring and early summer, males deliberately attempt to break into a female's territory by digging long, straight tunnels in the hope of intersecting a female's tunnel. Once he's found a female in estrus, copulation—as with the shrews—is perfunctory and brief and the male retreats to his own tunnel network. He will, in all likelihood, never even meet any of his offspring.

As with their fellow insectivores the shrews, moles find their prey mainly through smell and touch. To help them, their snouts are covered in thousands of tiny protuberances called Eimer's organs. Each of these tiny bumps is a separate organ, extremely sensitive to texture and touch.

Although it's impossible for us to know just how a mole interprets the information transmitted from these organs, it may provide these creatures with a textural picture of their environment as rich as our own vision or the sonar images of whales.

The star-nosed mole has an unusual snout ending in 22 radiating tentacles covered in Eimer's organs. It looks as if it has had a tiny sea anemone grafted to its nose. This arrangement, combined with eyes little larger than a pinhead, makes for a bizarre-looking animal. All moles are good swimmers, but the star-nosed mole is truly aquatic and spends much of its time under water. It can hold its breath for three minutes and dives a metre (3 feet) deep—often beneath ice in winter.

When a star-nosed mole is tunneling, the tentacles of the nose are folded away, but when it enters water, they open like the petals of a flower. As the mole swims along the bottom, it uses its highly flexible nose to probe for insect larvae, snails, and other invertebrates. Some biologists have postulated that, in addition to being a tactile organ, the star nose may be capable of detecting disturbances in electrical fields created by aquatic animals in the same way that electric eels and the elephant-nosed fishes find their prey.

The insectivores are the most primitive of placental mammals. They have evolved some extreme specializations, but their lives have probably changed very little from those of the first mammals: They are born, they search for food, they mate. The female gives birth to two or three litters and suckles them. Many die a scant 12 months after their birth.

A whale can lift 10 million times the shrew's weight clear of the ocean in a breach; the beaver builds a dam to protect and feed itself, perhaps without ever knowing why; human beings launch artificial satellites to communicate with others of their species whom they may never meet. These are the marvelous and extreme adaptations that impress us when we study the class Mammalia. But the works and bodies of all these animals have grown from the stem we still see in the insectivores, mammalian life distilled.

Dividing
the Spoils

Rodents hardly seem like a dominant animal group. They are not large as mammals go, nor do they form conspicuous features of the landscape like bison herds or flocks of snow geese. On the contrary, what energy rodents can spare from food gathering is spent in concealment. Their lives are ruled by efforts to avoid a legion of predators, and most rodents have made an art of remaining out of sight.

But if numbers are any measure of success, the meek have already inherited the earth. Since the evolution of the first rodents some 50 million years ago, very few rodent species have become extinct. Today, over half of all mammalian species are rodents, and these species are often populous. A hectare (2.5 acres) of tundra may be home to thousands of lemmings at the height of summer. At the turn of the century, a single prairie dog town in Texas was estimated to contain 400 million inhabitants. It covered nearly 65 000 square kilometres (25 000 square miles).

Being both small and abundant has made rodents the daily bread of North American predators. If it hunts for a living and has fewer than six legs, it probably feeds on rodents. Even shrews, mammals a fraction of their size, will attack and eat voles and mice. Few rodents have the

ABOVE: *The groundhog, like most marmots, is a true hibernator. Asleep in its underground burrow, its body temperature drops to a few degrees above freezing and its respiration slows to one breath every six minutes. Its heart rate may slow from 100 to just 4 beats per minute.*

LEFT: *Tree squirrels, such as this white phase of the eastern gray squirrel, are often cited for superior memories because of a supposed ability to find about 85 percent of the nuts they have buried. To test this theory, scientists buried nuts within a squirrel's territory. The squirrels found the planted nuts at about the same rate as the nuts they had buried themselves, so it seems the squirrels don't remember where the nuts are at all. They find them by smell.*

opportunity to die of old age. One of the most common and widespread, the deer mouse, lives up to eight years in captivity but in the wild 95 percent of them are eaten every year.

Rodents thrive under this continual assault because of their fecundity. A few, such as the porcupine, bear only a single offspring at a time, but the majority give birth to four or five young, and litters of up to 11 are not uncommon. The 13-lined ground squirrel can have up to 13 young in a single litter, and if they don't survive, the mother can give birth again the same year. Most mice and voles are capable of having three or four litters a year. Where the climate is mild and food abundant, a female deer mouse can have as many as 14 litters in a year. With a gestation period of three or four weeks, she may be almost continuously pregnant.

Chisel-like incisors characterize all of the rodents. These teeth grow throughout the animal's life, but constant use wears them down. The lower and upper pairs scrape against each other like a chisel on a whetstone. They are driven by bulging jaw muscles, which is what gives many rodents the chubby-cheeked look that human beings find so appealing. While they're designed for cutting down plants, cracking seeds and nuts, and scraping bark from trees and branches, these incisors can also give a predator pause. Most hunters try to steal up on larger rodents such as marmots and beavers to avoid being bitten.

Rodents generally protect themselves and their offspring not by fighting, but by nesting and living in burrows, and are nervous when they're out in the open. Although not as thoroughly adapted to underground life as the moles (rodents still forage above ground and most have excellent vision), almost all have compact, heavily clawed limbs and some of them dig as deep as the soil will permit. Prairie dogs dig some of the deepest and most elaborate burrow systems of any rodent, and their holes drop

RIGHT: *Despite a low reproductive rate for a rodent, porcupines are extremely successful. They enjoy a high survival rate and may live up to seven or eight years. Their longevity stems directly from the more than 30 000 quills covering their bodies. A predator's only chance to kill the porcupine is to attack the face or belly, the only parts of its body that are unprotected.*

almost vertically some 5 metres (16 feet) from the surface before leveling off. Their burrows interconnect with those of their neighbors' to form extensive colonies, so even predators able to follow them down their holes, such as weasels and ferrets, may have a hard time finding their quarry in the maze of tunnels.

Marmots are the largest members of the squirrel family, which includes the tree squirrels, chipmunks, ground squirrels, and marmots. The biggest of all, the hoary marmot, can weigh up to 14 kilograms (30 pounds). Marmots live in the mountains of western North America, just above the tree line. They need open meadows in which to graze on their principal foods—leaves, berries, flowers, and grasses—although they do eat some insects and the eggs of ground-nesting birds when they can find them. Marmots never stray far from their burrows, which often extend beneath talus slopes. These boulder piles are their only guaranteed protection from grizzly bears, which often try to dig them out of their homes.

The rarest of all rodents is the Vancouver Island marmot, of which there may be fewer than 70 remaining. No one knows why these animals are dying out, but the species' only hope may be for researchers to

ABOVE: *The red squirrel is highly arboreal, scampering through the trees with a speed matched only by the marten, one of its chief predators. The squirrel's tail not only helps it to balance when running along branches or telephone wires, but the drag created by the many long hairs also slows its descent if it falls. Many squirrels also use their tails as parasols in hot weather.*

LEFT: *Hoary marmots forage in alpine meadows from northern Washington to Alaska and their grizzled fur camouflages them against the granitic boulders common throughout their range. Boulder piles offer the marmots their best protection from grizzly bears, which may dig them out of burrows in soil, often when the marmot is still deep in hibernation.*

capture all the remaining marmots for a zoo breeding program and eventual relocation.

The groundhog, or woodchuck, is a more adaptable lowland marmot which will inhabit any open area where there is enough food. It has benefited greatly from the wholesale deforestation of eastern North America and the elimination of most of its predators. Today, its range is shaped roughly like a bell. The closed top is at the Alaska/Yukon border and the mouth covers almost the whole east coast of North America.

Marmots are deep hibernators, relying on ample fat stores to see them through the winter. Alaskan and hoary marmots living in the mountains of Alaska and northern British Columbia enter a profound sleep that may last from September to April. During this time, their body temperature drops to as low as 4°C (40°F) and their heart may beat only four times a minute.

With the exception of the groundhog, marmots are quite social, the males keeping harems of females within their territories. "Keeping" is perhaps misleading. The females could just as easily be seen to be retaining the male, who defends the territory from all other males. It is to the male's advantage to keep as many females in his harem as possible; the more females he mates with, the more offspring he produces. The female benefits from keeping the harem size as small as possible. Her litter size and the health of her young are improved by having to share food and other resources with fewer families. For this reason, females within the same harem are often less than friendly to one another. The more dominant females harass their inferiors, and will sometimes even plug their burrows with dirt in an attempt to suffocate them and their offspring. Often, the harems are matriarchies, made up of daughters and sisters. Rivalry between related members is reduced because they share many of

RIGHT: *The arctic ground squirrel is the most northerly of North America's ground squirrels. They live in Alaska, Yukon, and northern British Columbia west to Hudson Bay. They often choose banks near bodies of water or subalpine meadows in which to dig their burrows. They cannot dig through permafrost and this is probably what sets the northern limit of their range.*

the same genes. Hurting a relative is counterproductive to the gene's prime dictate: Produce as many copies of one's self as possible.

Marmot kinship extends to giving a warning whistle whenever they spot potential predators. Why animals give alarm calls has been the subject of intense scientific debate, and every year scores of scientific papers are published on the subject. On first consideration, a warning of some kind seems only sensible. If one marmot sees a predator and raises the alarm, it may save one of its fellow marmots. But if the marmot it saves is competing for food and territory, why should the alarmist want to help it? Furthermore, some experiments have shown that a warning call may draw the attention of the predator to the individual raising the alarm, putting it at greater risk. Most explanations hinge on the idea that the marmots nearest to the whistle blower are likely to be relatives, sharing many of its genes.

Ground squirrels exhibit varying degrees of sociability. Adult Columbian, Franklin's, and Richardson's ground squirrels all dig and defend individual burrows and only live together as juveniles. Still, a shortage of prime burrowing and foraging areas may bring them close to one another, with their burrow entrances no farther apart than those in a prairie dog colony. Female arctic ground squirrels (which are the largest of the ground squirrels, weighing up to 900 grams, or 2 pounds) live in harems much like those of the marmots.

The prairie dogs are the most social of all ground squirrels and occupy huge towns or colonies. They live on the grasslands and, like the bison, were numerous before deliberate human efforts to reduce their numbers. Unlike other ground-dwelling squirrels which do their best to conceal the entrances to their burrows, prairie dogs leave distinctive cones around some of their holes, making them look like little volcanos. The rim benefits the colony in a number of ways: A prairie dog can stand on it to improve its view of the surrounding territory. It keeps water from running into the burrow when the low-lying areas they inhabit are flooded. Most importantly, the rim helps to ventilate the tunnels, which can be almost 30 metres (100 feet) long. All prairie dog tunnels have at least two entrances, but as they are almost always dug into flat ground, the stale air would just lie in them. Building a crater around one of the entrances raises it a little higher than the other—usually just about 15 centimetres (6 inches), though sometimes it is twice that high. At ground level, the air is stagnant, but higher up there is a bit of a breeze. The movement of air across the crater rim creates a slight vacuum, pulling fresh air through the tunnel from the lower entrance.

Within each burrow system lives a group of females and a single male in a social group known as a coterie. As with marmots, the females are often closely related, but the coterie is a much more cooperative society than the harem. Once accepted, individuals are protected and welcomed as equals. The visible evidence of this society is the way prairie dogs greet each other, usually with a hug and a nose kiss, all made curiously human-looking by the prairie dog's upright stance.

Prairie dog mothers even nurse each other's young, although they may also suddenly decide to kill another's litter. Sometimes a male will keep more than one coterie, and in these cases the females of the coteries are likely to be hostile to each other. Litters consist of two to eight young and when the males are 12 to 14 months old, they leave the coterie to establish a territory of their own. They reach breeding age in their first or second year, and the average lifespan is three years, although captive animals have lived to eight.

The tree squirrels have abandoned a furtive, underground life for an arboreal one. Some of the tree squirrels openly scold and chatter at any

animal entering their territory and seem quite confident of their ability to elude predators as long as they are safely up a tree. Yet hawks, owls, and ravens are among the animals that prey upon them most heavily. The squirrels' agility and keen eyesight are their main defenses against these predators. In the trees, they can usually dodge a bird by darting to the other side of the trunk or branch. On the ground, they're much more vulnerable.

Tree squirrels have the largest brains relative to their bodies of all rodents. Evidence of their superior intelligence was once thought to be their supposed ability to remember the locations of their many nut and seed caches. It's true that tree squirrels do gather large numbers of nuts and seeds, usually climbing to them along branches, chewing through their stems and dropping them to the ground. The squirrel stores them either in holes they have dug or in the forks of trees. But biologists now know that squirrels find these caches with their sensitive noses. A squirrel can smell a nut through 30 centimetres (one foot) of snow. Even so, it misplaces about a fifth of its food. The trees benefit by having these lost seeds buried much farther afield than if they had just fallen.

Two species, the northern and southern flying squirrels, avoid competition by sleeping all day when other tree squirrels are foraging, and consequently very few people ever see flying squirrels. They emerge only after dark, when their huge eyes give them excellent night vision. As they launch themselves from a limb or tree trunk, a flap of skin running from wrist to ankle acts as a wing and allows them to glide at a 30- or 40-degree angle— usually to a neighboring tree. This is much more efficient than running down and up the trunks, and provides a quick escape from owls, their major predator. They even have some ability to turn and dip as they glide.

But the same flap of skin that enables the flying squirrel to glide is a

RIGHT: *The deer mouse is one of the most widespread rodents in North America, ranging from Baja to Labrador to the Yukon. Only the beaver and the porcupine have greater ranges. The deer mouse owes much of its success to its variability: There are at least 100 subspecies, and throughout much of its range it is the most common mammal.*

ABOVE: *Distinguishing one chipmunk from the 21 other species in North America is not a job for amateurs. Living chipmunks are often identified by their calls. Chipmunks in a less vocal mood often have to be distinguished by subtle variations in their striping patterns, which are thought to camouflage them from aerial predators. Taxonomists sometimes have to resort to examination of the penile bones for a positive identification. The chipmunk pictured here is a least chipmunk.*

hindrance on the ground, and they're very slow runners. Consequently, they spend almost all of their time in the trees.

In form, the 20-odd species of North American chipmunks closely resemble the tree squirrels, but the chipmunk's distinctive stripes serve to break up the body's outline as seen from above, making it harder for raptors to spot. This is particularly useful as chipmunks spend more of their time on the ground, foraging for bulbs, seeds, fruits, fungi, flowers, and nuts. Like tree squirrels, they cache many of these foods for the winter. In summer, they also eat insects, worms, slugs, and snails, and have even been known to attack and eat mice and birds.

Chipmunks nest in burrows. These can be under a metre (3 feet) long in the case of the yellow pine chipmunk to four times that length for the eastern chipmunk, but all are of relatively simple construction. As they dig, chipmunks use their cheek pouches to remove the dirt from their tunnels and scatter it some distance away from the entrance, rather like escaping prisoners hoping to conceal their diggings.

Neither tree squirrels nor chipmunks are true hibernators. Their high metabolisms make it difficult for them to store enough fat to sleep through winter. Tree squirrels remain active throughout the cold season, drawing on their food caches for sustenance. Many chipmunks enter a light torpor, rousing periodically to raid their caches and to drink. They have been seen breaking icicles from branches and sucking on them like lollipops.

The 18 species of pocket gophers are probably more thoroughly adapted to life underground than any other mammal. Like moles, they have powerful claws for digging, eyes that can shut tightly to keep out dirt, and hair that lies as easily in one direction as another. Gophers also have a few digging tricks that moles don't have: Their lips close behind their teeth so that they can use their incisors to chew their way through dense soils or pick out particularly stubborn rocks without swallowing dirt. They look a little like miniature bison with their tiny pelvises. This may make it easier for them to turn in their tight burrows.

Unlike moles, pocket gophers are herbivores, feeding mainly on the roots, stems, tubers and other subterranean parts of plants. The pocket gopher is named for the fur-lined external pouches that reach from their cheeks to their shoulders. These they fill with bits of vegetable matter they have cut from plants and transport them to the side tunnels and chambers where they store their food. They turn the pockets inside out to empty them.

Pocket gophers do not hibernate, retreating to their deeper tunnels to keep warm. They continue to feed throughout the winter, and deliberately plug their tunnel entrances to retain heat and moisture. To breathe, they must rely on the oxygen that permeates the soil, and so cannot live in clay or very wet soils. Many species have limited ranges, but as a group they are scattered throughout the central and western United States and the southern half of the Prairie provinces. They do not live in the tundra or boreal forests.

Male pocket gophers find a mate by digging in a straight line until they cross into a female's territory. Their litters are small for a rodent—averaging only three—but because of the gopher's well-protected environment, their survival rate is quite high. Many gophers are probably killed when they leave their mother's tunnels to seek a territory of their own. Outside of the mating season, which is usually spring to early summer, pocket gophers encountering each other will fight.

Living at ground level, the lemmings and voles rely on grasses and other vegetation to conceal their movements from predators. Voles and lemmings are a little larger and stouter than mice and have shorter tails, ears, and legs. Their fur is often shaggier and coarser. But in general, if one were to scurry across your path in the wild, the impression would be of a mouse. In fact, the North American term "field mouse" is synonymous with vole.

Even by rodent standards, the voles and lemmings are a highly successful group. There are 70 species of lemmings and voles in North America, and there is little observable difference from one to the next. The obvious question is, what is one species doing that another couldn't do just as well?

Voles and lemmings are all nocturnal herbivores, but they divide the plants available in quite specific ways. The tundra and boreal forests are home to paired species of voles, one of which takes the higher, hummocky ground to feed on grasses and the bark and leaves of willows. The other feeds on the sedges and mosses growing in the soggier, lower zone.

Farther south, most voles live in forests, but a few prefer open meadows. In general, the voles living in open meadows have more efficient digestive systems. Not only their incisors but also their molars grow throughout their lives. If they didn't, their teeth would wear down to nothing within weeks because the grasses they eat accumulate silica and are very abrasive. The intestines of meadow voles are long for their body size, with special chambers for fermenting bacteria that help them to extract every calorie from the nutrient-poor grasses. Forest voles are able to supplement their diet with berries and seeds, which have many more calories, and so their intestines are shorter.

Most lemmings and voles live in a series of burrows or rock crevices during summer and travel from one to another using runways they wear through the grass. They remain active throughout the winter, usually tunneling through the snow at ground level. There's very little greenery at this time of year, so they dig down to the roots and rhizomes of grasses and feed on them. Others survive on bark and seeds. Only a few species stockpile food to sustain them through the winter.

Lemmings are renowned for their fecundity. A female can become pregnant while still nursing her last brood, and litters can be as large as eight. With such reproductive capacities, the populations of lemmings can explode from a few individuals per hectare to thousands over the course of several summers. When their numbers peak, every carnivore and

raptor feasts on the lemmings, which seem to be everywhere. In the north, where lemmings live, most predators are utterly dependent upon them for their survival. But as they don't have nearly the reproductive rates of their prey, it may take a few years for their numbers to build to the point where they can seriously affect lemming populations. Before predator numbers catch up, lemmings may almost strip the land of their preferred plants.

Lemming suicide is one of ecology's most persistent myths. It was once thought that lemmings would periodically take it upon themselves to eliminate population excesses by plunging en masse into the sea. Later, these drownings were assumed to be migrations that had taken a wrong turn. Today, most biologists even doubt the idea of coordinated mass migrations. Lemmings are so abundant at times that large numbers of them can probably be seen running in the same direction at once. But experiments tracking individual lemmings have found that they do not move in one direction for any length of time. On the contrary, voles and lemmings have quite strong homing instincts.

Just as paired species of voles may divide the tundra, kangaroo rats (which are not rats at all, but relatives of the pocket mice) and deer mice share the deserts of Mexico and the American Southwest. At night, they forage for the same seeds, roots, leaves, and insects. Both are hunted by the same predators—snakes, owls, ferrets, and foxes. But experiments using video cameras and fine sand to track the movements of these rodents have shown that the larger and more aggressive Ord's kangaroo rat, which prefers open areas where it can hop along rapidly on its hind legs, will drive any deer mice it encounters into covered areas.

Kangaroo rats have large eyes placed high on their heads for spotting flying predators. Unlike other rodents, their big eardrums are designed for detecting low-frequency sounds—the wingbeats of an approaching owl

LEFT: *The 13-lined ground squirrel is a true omnivore. Like most ground squirrels, it feeds mainly on seeds but it also eats insects, carrion, and even mice and shrews during the late fall. Its varied diet allows it to put on a heavy fat layer in preparation for its deep winter hibernation.*

ABOVE: *Porcupine litters usually consist of a single kit but occasionally twins are born. To avoid injury to the mother, the young emerge head-first encased in a placental sack. The quills are quite soft at birth but within half an hour they harden. In two hours, the young porcupine can raise them if threatened.*

or the air pressure wave that precedes a snake's strike. By carefully timing one of their 2-metre (6-foot) hops, they can evade these predators, but a kangaroo rat may also choose to blind an attacker by turning its back and running on the spot, spraying a cloud of sand and dirt into its eyes.

Porcupines are large rodents, second in size only to the beaver, and a big male can weigh 14 kilograms (30 pounds). They exploit a different food source than most other rodents—mainly leaves and bark. In summer they feed mostly at ground level, but in winter they climb deciduous trees to feed on the buds and bark. They also eat the needles of conifers.

The predators of this slow-moving animal would be many were it not for the modified hairs, or quills, covering all but its face and belly. Any animal touching one will find the barbed point instantly embedded in its own skin, while the root of the quill detaches from the porcupine.

Because of the scales covering the quill, any movement of the shaft causes it to work its way deeper into the victim's flesh. On rare occasions, the quills reach a vital organ and kill the animal. More often, inexperienced lynxes, bobcats, wolves, or bears that have made the mistake of attacking a porcupine die of infection, or starve to death because their muzzles are so full of quills they can't eat. Those that survive the experience rarely repeat the mistake.

A few animals specialize in feeding on porcupines. The fisher attacks the porcupine's face, biting it until it bleeds to death or dies of shock. It then gets a paw under the body, flips it over, and chews into the belly, eventually eating all but the skin. Cougars are reported to flip the porcupine over first and kill it with a bite to the stomach.

The question that inevitably comes to mind is how do such well-protected animals get close enough to one another to mate? Surprisingly, porcupines do not copulate belly to belly. The male begins courtship by repeatedly grunting and rearing on his hind legs. His erection comes not from the usual engorgement of the penis with blood as in most mammals, but by the action of an intricate system of muscles. He sprays the female with urine and chases her about, grunting and wrestling with her. Eventually, she stretches out her hind legs and folds her tail (normally one of the most dangerous parts of the porcupine's armor) over her back, exposing a quill-free surface for the male, who inserts his penis. He can't actually grasp her, so his forepaws hang free during mating.

Copulation is brief and, as with many mammals, part of the male's ejaculate solidifies into a vaginal plug, preventing other males from fertilizing the female. Porcupines are much less prolific than other rodents. After a lengthy pregnancy of 30 or 31 weeks, the female usually gives birth to just a single kit. Although it is born with eyes open and can feed on vegetation within a week, the mother is nonetheless attentive, and nurses her young throughout the summer. Consistent with a caring parent and a long life (porcupines can live up to 10 years in captivity), porcupines enjoy a high infant survival rate. This contrasts sharply with the typical rodent strategy of producing many, short-lived offspring, but it seems to have paid off for porcupines. They are one of the more common large animals in the mixed boreal, hardwood, and coniferous forests of North America.

The beaver is another rodent that feeds on leaves and bark, but it acquires these foods in a completely different manner. In the course of its evolution, the beaver has invented an ingenious way to maintain a supply of food year round as well as to protect itself from the predators that plague other rodents. It builds one of the largest of all animal structures, the beaver dam. Not surprisingly, it is itself a giant among rodents,

weighing up to 28 kilograms (60 pounds). Only the capybara of South America—another aquatic rodent which looks a little like a giant, long-legged guinea pig—is larger.

Beavers have such large incisors that they can gnaw through the trunks of trees up to 85 centimetres (33 inches) in diameter. Cutting down trees probably began as merely gnawing at the inner bark, which beavers are able to digest. An overzealous beaver, gnawing through a tree trunk and felling it, would have immediately benefited by being able to reach the tender branches and leaves farther up the trunk. Dam-building likely began by accident when some of these trees fell across a stream.

But the process of building a dam has become much more refined than the happenstance of that first rodent engineer. Beavers are now experts at eyeing the terrain and deciding where building a dam will achieve the desired effect. Once they have chosen a suitable site, they grasp sticks in their mouths and drive them upright into the streambed. They lay branches across these piles and weigh them down with rocks rolled from the banks of the stream. As the water starts to rise, they seal the whole construct with mud and clods of earth dug from the banks. Beavers are very particular in the placement of each branch and stick, often adjusting its position several times before being satisfied. They pile more and more branches and mud on the downstream side of the dam to counter the pressure of water rising on the other side, until the whole structure may weigh many tonnes.

Damming streams provides beavers with ponds where they are safe from most predators. With its webbed feet and paddlelike tail, a beaver can easily outswim coyotes, lynxes, foxes, bears, or any of the other carnivores that would quickly make a meal of it were they to catch it on dry land. The pond also serves as a cold storage for winter food. Before the

Left: *The muskrat's way of life is more like a beaver's than like any other rodent's, but it is in fact the largest member of the family that includes mice, lemmings, and hamsters. The muskrat is really a giant vole, outweighing any of its cousins by at least 10 times. This is probably necessary for it to maintain its body heat in the water, where it spends much of its time.*

leaves turn in autumn, beavers cut down saplings and drag them to the pond they have made. There, the cold water keeps them fresh and green through the winter. Aquatic plants growing in the pond provide a further food source.

Perhaps the beaver's most ingenious construct is not the dam, but the beaver lodge—the hollowed-out mound of sticks and mud in which the beaver lives, sleeps, and bears its young. Before it builds the dam, the beaver usually begins its work by digging into a bank of the stream. At some point, it begins burrowing upward. Just before it breaks through the ceiling of the burrow and into the open air, it stops digging. Then it piles the ground over its excavation with sticks and mud, much in the same way it builds a dam. The beaver then returns to its burrow and digs upward until it reaches the bottom of the pile of sticks, which now serves as the roof of its lodge. It removes some of the sticks from the bottom of the pile to further enlarge the space. Eventually, water behind the dam rises to a level where the entrance to the lodge is completely submerged and the beavers can come and go underwater.

Beavers may also build their lodges on the dam itself or, ideally, on an island in their ponds, making a moat out of it. Other beavers den in simple riverbank burrows. Entrances to lodges and dens are underwater, denying access to most predators. A bear may tear apart a beaver lodge to get at the beavers, but it will usually take enough time that the beavers can slip out through the entrance before the bear has gotten through. In winter, when the dome of mud and sticks is frozen solid, even an animal as strong as a bear may have trouble pulling it apart. Their ability to alter their habitat has earned the beaver one of the most extensive ranges of any mammal in North America. They live from the Rio Grande to the far north, right across the continent.

The muskrat also relies on water for protection from its enemies, and shares many of the beaver's behaviors. It looks essentially like a tiny beaver, weighing at most 1.4 kilograms (3 pounds) but having a naked tail that is only slightly flattened, from side to side. It has the same thick underfur and glossy outer coat that insulate the beaver from the coldest water and, like the beaver, the muskrat can also hold its breath for as long as 15 minutes. It, too, lives in homes with underwater entrances. Muskrats do not build dams, but will construct lodges out of cattails or

LEFT: *The fox squirrel is the largest of the tree squirrels, weighing up to 1.4 kilograms (3 pounds). It lives throughout the eastern half of the United States. Fox squirrels prefer to nest in tree cavities, but will make do with the crotch of a limb lined with leaves either as a den or to cache hickory nuts, their favorite food.*

bulrushes in marshes or other bodies of water where there are no banks to burrow into. They also reinforce their homes with mud.

The muskrat supplements the beaver diet with some animal protein, eating frogs, crayfish, mollusks and even the young of waterfowl. In turn, it is preyed on by a wide variety of raptors and carnivores. Mink are their worst enemies because they swim well and can follow the muskrat into its lodge. Although they can live for up to 10 years in captivity, few muskrats survive beyond three or four years of age before they are eaten.

Despite their superficial similarities to beavers, muskrats are actually giant, aquatic voles. They are one of the most successful of the rodents, living in abundance in ponds and lakes from Mexico to the Arctic Circle.

Most of us give little thought to rodents until they interfere in some way with our own lives, and at such times we are not inclined to look upon them kindly. Mice and rats rob us of the grains we stockpile, squirrels and chipmunks gnaw through walls and wiring in our homes, beavers flood farmers' fields, and gopher mounds damage agricultural machinery. In short, we think of them as pests.

Most of the damage they wreak stems from the rodents' strongest assets—their abilities to stockpile foods and to tunnel. The irony is that in exercising these skills, they often derail our endeavors in the equivalent sciences of agriculture and engineering.

RIGHT: *Hoary marmots frequently wrestle, rearing on their hind legs to push at one another. Although no one knows the purpose of these shoving matches, juveniles engage in them more often than do adults. The bouts may be a form of play that reduces tension among individuals sharing pasture and other resources.*

Rabbits Run

In the Aesop's fable of the Hares and the Frogs, the hares gathered to lament their lot in life, devoured daily by men, dogs, birds, and beasts of prey. Seeing no hope, they one and all determined to end their miserable lives by rushing into a nearby pond and drowning themselves. But as they dashed headlong toward the pool, the frogs sunning themselves on its banks were startled and leapt into the water to hide themselves in its depths. Seeing this, one wise old hare cried out for the others to stop and take heart, for there were, after all, creatures more timid and put upon than themselves.

The lagomorphs—pikas, rabbits, and hares—could be forgiven for having something of a persecution complex. Like their cousins, the rodents, they are besieged by predators. Once considered a suborder of the rodents, they share their ever-growing incisors and their keen vision, hearing, and sense of smell. Like the rodents, they have answered their predators with numbers. Rabbits are among the most prolific of mammals.

But there are important differences between the two groups: The lagomorphs have not one, but two pairs of upper and lower incisors. They are vegetarians, eating grasses and other green vegetation. A number of

ABOVE: *Several small mammals whose ranges extend into high latitudes have coats that change color with the seasons. The snowshoe hare is shown here in its brown summer phase. The seasonal molts that produce the color changes are triggered not by temperature, but day length. Only the outer guard hairs are replaced. The underfur remains brown throughout the year.*

LEFT: *The marsh rabbit is plagued by the usual array of predators, but it often deals with them in a novel way: It takes to the water, either swimming away rapidly or hiding among aquatic plants with only the tip of its nose above water. On land, the marsh rabbit can be identified by its rather short, rounded ears.*

RIGHT: *This American pika's alarm call, a high, penetrating "eek," is easy to hear but harder to pinpoint. The sound bounces around the boulder piles where the pika lives, and the pika's granite-colored pelt also makes it difficult to pick out from a rocky background. Like many burrowing animals, after their initial dash for cover pikas will often return to their burrow entrances to get a second peek at human intruders.*

species have evolved from scurrying between burrows to out-and-out sprinting to escape predators.

The pika is the most rodentlike of the order. In North America there are only two species, and they both require one very particular habitat: the talus slopes high in the western mountains. Talus is the jumble of stones or boulders that accumulates below rock faces. Where the rocks are of the correct size and border on a meadow, pikas live in the spaces between rocks and in burrows they dig beneath them. They leave the safety of the talus only to forage for the grasses and flowers that are their main food—hence their other common name, rock rabbit.

They don't look much like rabbits. A large pika is about the size of a guinea pig. With their small, rounded ears, tailless rumps, and brown to gray fur, they more resemble the rock part of their name. Hikers, the only people to commonly see pikas, would probably walk past them unaware, but when a pika sees an intruder it gives a high, warning "eek." Even then, these animals are so well camouflaged that spotting them among the rocks is difficult.

Aside from the protection from predators afforded by the talus, pikas have another reason for this strict habitat requirement. They cannot tolerate heat, and will die if exposed to temperatures much above that of their own bodies for very long. High in the mountains, the air inside their talus dens remains cool. Pikas seem to have no trouble surviving the cold alpine winters, and they spend much of it under deep snow.

During the May and June breeding season, pikas are quite social, almost colonial animals. About 30 days after mating, two to six pink, blind pikas are born to the female. They grow rapidly, reaching full size in only 40 to 50 days. But even earlier, they leave their natal burrows to find their own territory.

With their specific habitat needs, suitable territories are in short supply, and a change comes over the pikas as the summer progresses. They start to defend their territories and at the same time, they devote more and more of their time to cutting flowers, grasses, and just about any green plant with their incisors and then carrying them back to a flat boulder within their territory. They move or turn the pile several times a day to keep it in direct sunlight so that it dries out. While a pika is out foraging, its neighbors will often try to raid its hay pile. Owners returning to catch these bandits in the act will chase them from their territory.

The pikas store the dried vegetation underground in hay piles up to

35 litres (one bushel) in volume. They do not hibernate, but draw on their hay piles to sustain them through the long alpine winter. If it's lucky, a pika will live to see five to seven such winters.

Like all the lagomorphs, pikas eat some of their own fecal pellets to extract certain vitamins and as many calories as possible from them. It is the equivalent of the ruminants chewing their cud. Pikas excrete two kinds of pellets: those that have made only one trip through their digestive systems, and those that have been through twice. Pikas and rabbits are able to pluck those that are on their first time through directly from their anuses. The practice is known as refection.

In sharp contrast to the pika, hares live where there is nowhere to hide, on the open prairie, deserts, and tundra. To escape their many predators, hares make a run for it and their oversized rear legs give them the speed they need. The white-tailed jackrabbit, which lives in the grasslands and central plains of the continent, can sprint at 65 kilometres (40 miles) per hour. Its strong legs enable it to make such sharp course changes that even a predator with a higher top speed, such as a coyote, may fail to catch it.

Rabbits are also fast runners, but are distinct from hares in a number of ways. The young of hares are born in the open and with their eyes open.

ABOVE: *Arctic hares have a number of adaptations to the extreme winter cold of their range, which extends to the northern tip of Ellesmere Island. They are large, with comparatively blunt paws and ears, and well-furred feet. With their long incisors they can tweeze tiny plants from between stones and rock crevices. They have been seen walking upright on their hind legs.*

LEFT: *You wouldn't guess it from this particular rabbit's stern expression, but Nuttall's cottontail is rather shy, even by rabbit standards, and bolts for heavy brush at its first whiff of trouble. Nuttall's cottontail lives in alpine meadows and the high plains of the western United States, southern Alberta, and southern Saskatchewan hence its other common name, the mountain cottontail.*

They can be up and running within minutes of birth. Rabbits are born blind and helpless in a shallow hole the mother has dug for them and lined with her own fur. (It is the European rabbits that live and nest in the permanent, communal warrens of *Watership Down* fame.) Most rabbits are slower runners than hares. Their hind legs are not nearly so large, and many will run for cover when danger threatens. The name "jackrabbit" is often applied to hares, but is a misnomer, while the Belgian hare is actually a species of rabbit. The defining differences are in the structure of the skull, and are not readily visible.

Most hares and rabbits stand still on first becoming aware of a predator in the hope that they won't be seen at all. But if a hunter approaches within eight or nine metres (26 to 30 feet), it triggers the animal's flight response. Accordingly, most predators try to take their rabbits by surprise, but it isn't easy. Their eyes, located far to the sides of their heads, give rabbits and hares nearly 360-degree vision. The most distinctive feature of the rabbits and hares, their large, upright ears, are also extremely sensitive. Hunting these animals might be compared to trying to sneak up on an antenna continuously tuned to the predator band.

And yet, almost all hares and rabbits die by predation. Because they are active mainly from dusk until dawn, owls are a major killer. Eagles, lynxes, bobcats, foxes, coyotes, wolves, and weasels also take their share of adults. Ravens and crows kill many of the young. In captivity, cottontail rabbits have lived to nine years, but in the wild, few live to see their second birthday. Wild hares fare a little better, usually surviving three to five years.

Even so, predators cannot keep up with the rabbits, which are not the symbol of fertility for nothing. The eastern cottontail, which is common throughout the eastern United States, gives birth to litters of 2 to 12 young after a gestation period of 28 days. It can have up to seven

LEFT: *Pikas do not hibernate, but collect large piles of flowers and leafy plants to see them through the winter. After they cut down the plants, they find an exposed rock on which to dry them, in direct sunlight when possible, before taking them underground for storage.*

litters in a single year, and the young are fertile within two or three months of birth. Of the mammals, only some of the small rodents and shrews surpass this rate of production. Hares have fewer and smaller litters, but enjoy a much higher rate of survival, both as juveniles and as adults.

During spring courtship, male hares and rabbits leap around the females, urinating on them while thumping the ground with their hind feet. For their part, the larger females sometimes rear up and box the males, cuffing them with their forepaws. This bruising is probably a test of the male's fitness and his determination to mate. His urine may also give the female clues as to his physiological readiness to mate—although for the rabbit family, this would seem to be redundant information. Copulation is brief, and afterward the pair separates. The male offers no help in rearing the young.

Rabbits and hares have managed to exploit almost every type of habitat in North America, from the high arctic to the southern deserts, from swamps to alpine meadows. Although they vary in size, their basic body design changes little from one biome to another. They all have highly mobile lips which meet behind their incisors when their mouths are closed. Their cheek teeth grow continuously to compensate for wear from

the abrasive vegetation they often eat. Their small intestines are long, and a pouch called the cecum, at the junction with the large intestine, contains bacteria that aid digestion.

The arctic hare is the most northerly of the lagomorphs with a range extending well above the Arctic Circle. It is the largest of hares, reaching 6 kilograms (13 pounds) and has thicker limbs and shorter ears, all of which are helpful in conserving heat. To reach the twigs and roots of the low-growing tundra plants on which it feeds, it scrapes through snow and ice with its rather large claws. It also has longer incisors than other hares. The coat of the arctic hare is white in winter, gray in summer.

Arctic hares will eat carrion whenever they get the opportunity, particularly in winter when forage is scarce. To see several hares chewing at the body of a dead seal or fox must rank as one of nature's more unnerving sights. Like many arctic animals, they show very little fear of human beings in their rare encounters with them.

The snowshoe hare lives in boreal and mountain forests. It's also called the varying hare because, as with the arctic hare, its coat changes color with the seasons. In winter, it is white except for the tips of its large ears. During the summer, its coat is a mixture of browns and grays. (Only the outer, guard hairs are replaced; their underfur remains a constant brown.) The snowshoe hare has very broad paws with thick hair growing between the pads. With them, it can float on snow that leaves many of its predators floundering. Unfortunately for the snowshoe, its major predator, the lynx, has similarly large paws and is also undeterred by deep snow.

Unlike the jackrabbits, the snowshoe hare is quite gregarious outside of the mating season, and sometimes feeds in groups. During the summer, they eat just about any green vegetation they can reach, including

LEFT: *A snowshoe hare appears to offer thanks for living another day. Despite keen hearing, eyesight, and sense of smell to detect predators and large hind feet that allow them to run over the lightest snow, snowshoe hares almost never die of old age. Up to 40 percent of the population is eaten each winter.*

ferns. In winter, they survive on buds, twigs, and bark, as well as the leaves of evergreens.

As with all desert animals, the main difficulty for the black-tailed jackrabbit (a hare) is finding and retaining water. It lives in the deserts and prairies of the American southwest and during the heat of the day it dozes in the shade of rocks or shrubbery. At night it forages for grasses, herbs, twigs, bark, and mesquite. In true deserts, it gets almost all of its water from cactuses. The black-tail is probably the fastest of all the rabbit family, capable of speeds up to 80 kilometres (50 miles) per hour. Remarkably, there are still predators that can catch them—mainly raptors, coyotes, and swift foxes—and the average black-tailed jackrabbit lives less than a year.

The smallest of North American rabbits is the brush rabbit. It lives in the short sage and open grasslands of eastern Oregon, parts of Idaho, and Nevada and is the only North American rabbit to dig its own burrow. It weighs just 340 to 400 grams (12 to 14 ounces).

In the marshes of the southeastern United States lives the swamp rabbit. While most rabbits avoid water, swamp rabbits swim without hesitation, fording streams and rivers to escape their predators or reach new forage. On firm ground, they can run almost as fast as a jackrabbit, usually dodging their pursuers with a zigzag pattern. There are accounts of them hiding from predators submerged, with only their twitching noses above the surface of the water.

Male swamp rabbits establish dominance hierarchies, and when one intrudes on another's territory, they often fight viciously. They face each other standing on their hind feet and rake at each other with their teeth and claws. But the worst wounds are inflicted when they leap up and slash at each other with their hind claws. These fights can be to the death.

The lagomorphs have thoroughly exploited the land while being exploited themselves. An animal with the jackrabbit's keen senses, capable of running at highway speeds, would seem to be almost invulnerable to predation, and yet it is the single most important food for many predators. In the United States alone, human hunters shoot an estimated 50 million rabbits and hares every year—and this is just a fraction of the number taken by other animals. The predators of the world would soon starve if the hares, taking their cue from the fable of old, ever did commit to a suicide pact.

LEFT: *The collared pika has a more limited, northerly range than its relative, the American pika, and is found from southeast Alaska to the southwestern Northwest Territories. It's not much bigger than a guinea pig, but its compact, well-furred body allows it to remain active throughout the winter months. In fact, pikas tolerate extreme cold far better than they do heat, which is probably why they are found only in alpine habitat.*

ABOVE: *Unusual among the lagomorphs, arctic hares sometimes congregate in groups of hundreds. Gathering in this manner may afford the hares some of the same protection from predators that herding provides antelope or caribou. It may also help the hares to stay warm in the frigid climes where they live.*

Southern Invaders

About three million years ago, a chain of volcanoes arose between North and South America to create a land bridge, and something resembling a species exchange program began between the two continents. Not that animals rushed across the Panamanian isthmus in any determined way. Their ranges slowly expanded, probably over tens of thousands of years. Many of these animals fared better in their new homes than in their old, where they eventually became extinct. The porcupine evolved in South America but no longer lives there. Others, such as the raccoon, coati, and opossum, made the northern migration but still live on both continents.

But crossing the isthmus from South America and spreading northward—as far as Canada in some cases—meant adapting to a radically different climate. Perhaps it is for this reason that many of the most successful migrants eat a wide variety of foods, are adaptable, and can be quite intelligent.

Certainly, the members of the procyonid family fit this description. Raccoons, ringtails, and coatis will eat almost anything they can find, with diets similar to that of a black bear. The coati inhabits the woodlands

ABOVE: *The nine-banded armadillo is North America's only member of an order that also includes the anteaters and tree sloths of the New World. One of the names for this order, Edentata, means toothless. While this accurately describes some armadillos and anteaters, it's inappropriate to the nine-banded armadillo, which has spikelike molars.*

LEFT: *Unusual for a member of the raccoon family, the coati forages by day. The females are highly social, traveling in troupes of 6 to 25 juveniles and adult females. Males leave the troupe for a more solitary existence upon reaching sexual maturity. This is not an unusual mammalian social arrangement, employed by animals as diverse as sperm whales and ground squirrels.*

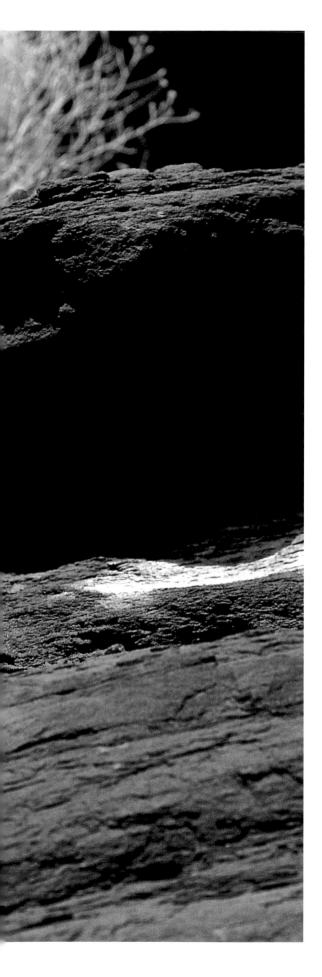

and scrubland of Mexico and the southwestern United States. Ringtails have penetrated as far north as the Oregon-California border in the west, and still seem to be increasing their range in the east where they have spread into Kansas, Arkansas, and Louisiana. The extermination of many of their larger predators has probably encouraged the expansion.

The ringtail resembles no other single animal. It has a bushy tail as long as the rest of its body, marked with even more distinct rings than a raccoon's. Its face is like a small fox's, but the eyes are large and very dark, suited to the ringtail's nocturnal foraging. The body is something like a squirrel's in form, but much larger. Adult males weigh a little over a kilogram (2.2 pounds). Like its relative the raccoon, the ringtail has dextrous, almost handlike front paws.

During the winter, it hunts for small rodents in the cliffs and rocky terrain that are its habitat. It may range over 100 hectares (250 acres) in search of prey. The ringtail has sharp carnassial teeth and is a skilled carnivore, delivering a fatal bite to the neck much in the manner of a cat. It often begins consuming its prey by the head. Like the fisher, retractile claws and a swiveling rear foot allow it to descend trees headfirst in pursuit of prey. When the weather warms, the ringtail's diet expands to include spiders, crickets, grasshoppers, lizards, scorpions, amphibians, eggs, and fruit.

There is some controversy as to whether ringtails live alone or in pairs, but the male sometimes assists in rearing the litter of two to four young by bringing food to them when they are three or four weeks old. More commonly, the female rears them on her own for four months. Ringtails were once called miner's cats because they were tamed by prospectors in order to rid their camps of vermin. Most live about five years in the wild, and many are killed by owls or bobcats.

LEFT: *The nocturnal ringtail spends its days in the hollow of a tree or sleeping in a rock crevice, and so it is unusual to see one moving about in full daylight as this one is. More carnivorous than raccoons or coatis, the ringtail hunts for almost any animal smaller than itself and kills with a quick pounce. Ringtails also eat a variety of plant matter, including corn, berries, nuts, and fruits.*

ABOVE: *Like its relatives, the ringtail and the raccoon, the coati fights when cornered. Its first reaction to most predators is to climb the nearest tree, but it has also been known to turn on its pursuers. The coati has a reputation for killing domestic dogs.*

RIGHT: *Touch is one of the raccoon's most highly developed senses. With their dextrous, almost handlike paws, they feel beneath logs and stones at the margins of streams and ponds for crayfish, frogs, salamanders, and snails. Folk wisdom has it that raccoons always wash their food before eating it, but more accurately, they are probably using water to separate shells from the flesh of their invertebrate prey.*

The coati looks like a raccoon weighing 4.5 to 11 kilograms (10 to 24 pounds), except that it has a much longer snout and often walks with its tail erect, exclamation mark fashion. The ring markings of its tail are less distinct, but like the raccoon it has a black face mask. Coatis are the most gregarious of the procyonids, and females and young forage in troops of up to 30—usually during the day—searching for seeds, fruits, and animals from crayfish to insects. At night, they sleep in trees.

Like most social animals, coatis are quite vocal, snorting, grunting, chattering, or screaming, depending upon their mood. The young entertain themselves by chasing each other up and down trees. In many ways, their behavior more resembles that of a primate than it does the raccoons. They have a reputation for fighting fiercely when cornered, and have been known to kill domestic dogs.

Males over three years old are solitary, but during the April mating season they ingratiate themselves with a troop by joining them in communal grooming and other submissive behaviors. Having infiltrated a troop, they eventually mate with most of the adult females. Copulation usually occurs in trees. The male is then expelled from the troop, as he has a tendency to kill young coatis. Although they are larger than females, a male

will be attacked by the troop if he attempts to approach it outside of the mating season. For a time, biologists mistakenly thought these exiled males to be a separate species, the coatimundi.

About 77 days after conception, the female leaves the troop to give birth to four to six young, usually in a rocky den or abandoned animal burrow. She returns to the troop with them five to six weeks after they are born. Coatis live up to 14 years in captivity, but their lifespan in the wild is probably considerably less.

Judged by either size or range, the raccoon is the most successful member of the procyonids. They live from South America to Hudson Bay, and one giant male in Colorado weighed 27 kilograms (almost 60 pounds). However, an individual just half this size is still big for a raccoon. Those living in southern Florida rarely exceed 4 kilograms (9 pounds). Like most animals, individuals living farthest from the equator tend to be the largest. Raccoons live in a wide range of habitats, from swamps to jungles to hardwood and coniferous forests, but they rarely stray far from a stream or other body of water.

Raccoons have also succeeded in penetrating our cities, where more of them live now than in the wild. Like most urban animals, they adapt readily to new food sources. Being nocturnal is a help. Although their eyesight is not particularly good during the day, they have strong night vision, and their keen noses lead them to garbage cans and compost heaps. Their claws can easily tear through garbage bags, but a raccoon's tiny paws are remarkably deft. Raccoons often make nuisances of themselves in campgrounds by raiding the tents and packs of careless campers for food. But in at least one campground on the San Juan Islands off the coast of Washington, rather than chew or tear through the nylon, they've learned to search backpack storage compartments by carefully unzipping them.

RIGHT: *Like many of the mammals that have spread into North America from the south, the Virginia opossum is not fully adapted to the harsh winters in the northern reaches of its range. The naked and somewhat ragged-looking ears of this one have probably been damaged by frostbite.*

The reason for the raccoon's dexterity is the large number of nerve endings in its paws. The part of its brain responsible for tactile sensation is correspondingly large and well developed for an animal its size. Raccoons are among the most intelligent of mammals, believed second only to the primates, whales, and possibly the dogs, to which they are distantly related. Many of their behaviors are learned, passed from mother to kit.

The name raccoon probably comes from an Algonquin word meaning "scratches with hands," and their paws are forever moving, usually probing for food. In addition to all the foods that ringtails and coatis feed upon, the raccoon also relishes slugs. Raccoons have a reputation for washing their food before eating it, but this is a myth that may have arisen from their habit of standing on the margins of streams or ponds and feeling beneath rocks and logs for crayfish, snails, fish, and other small, aquatic animals.

In winters in the northern parts of their range, raccoons often hole up in a high tree cavity or animal burrow for weeks at a time, but they are not true hibernators. They rely instead on accumulated fat stores—up to half their body weight—and their ingenuity to see them through the winter. On cold nights, they have been seen napping on bannisters and porch railings decorated with Christmas lights in an effort to keep warm. Urban raccoons will alternate among three or four dens, often moving into crawlspaces or attics to bear their one to seven young. The juveniles stay with their mothers for their first summer and winter, learning how to forage. Otherwise, they are asocial, but groups of up to a dozen have been known to share an attic or an abandoned cabin to conserve warmth.

Wild cats, wolves, and birds of prey all feed on raccoons, but most other carnivores are reluctant to attack them as they fight if cornered. People maintaining their distance from raccoons have nothing to fear from them, although in some places raccoons are among the more common carriers of rabies. Captive raccoons have lived up to 20 years, but their lives in the wild are only a fraction of this, perhaps 4 or 5 years.

Opossums have been only slightly less successful than the procyonids in expanding into North America. In South and Central America, there are some 47 species of opossum but only one, the Virginia opossum, has extended its range northward. With their small bodies—about the size of a house cat—and their naked paws and tails, opossums appear ill-equipped to survive a long, cold winter. The exposed parts of their bodies, particularly the tail, are subject to frostbite, which sometimes develops

LEFT: *Although eager to learn, this young raccoon hasn't quite got the knack of balancing on a thin branch. Like many intelligent animals, raccoons acquire many of their adaptive behaviors by imitating their mothers and siblings. Learned behaviors may require more time to develop than instinctive ones, but have the advantages of adaptability and improvement throughout the animal's life.*

ABOVE: *The opossum is North America's only marsupial, a group of animals considered to be more primitive than the placental mammals. This is partly because marsupials have some primitive skeletal features that are closer to those of reptiles than to other mammals, and also because marsupials often fare poorly when exposed to competition with placental mammals. The Virginia opossum is bucking the trend, however, and is still spreading northward while many mammals within its range are in decline.*

into gangrene. Having only immigrated from South America after the last ice age, they have no ability to hibernate or store food. And yet, they survive winters as far north as southern Ontario and British Columbia.

Prior to the arrival of Europeans, opossums lived no farther north than Virginia and Ohio. But they have thrived under the growth of agriculture and the elimination of their largest predators—wolves, cougars, and fishers. They were introduced to southern British Columbia as small game shortly after the Second World War.

Opossums eat many foods, including insects, lizards, amphibians, eggs, fruits and vegetables, carrion, and small mammals—even shrews and moles, which are shunned by many other predators because of the musk they exude. Opossums also eat snakes, and are almost immune to the bites of rattlers, water moccasins, and copperheads.

Like the raccoon, they have fleshy, naked paws. Their hind paws have an opposable toe, which they use, along with their prehensile tails, to grip tree limbs. They are excellent climbers. Opossums are very good at accumulating fat and, by fall, it may account for up to 30 percent of their body mass. They rely heavily upon this store to help sustain them until the spring in the northern part of their range.

Opossums are North America's only marsupials, and are among the most primitive of mammals. They have tiny brains—only one sixth the size of a raccoon of the same weight—but nonetheless seem highly adaptable. Their best-known defense against predators is to play dead. Finding itself in the jaws of a dog or other common predator, opossums usually lie completely still with their mouths open. Their heart and breathing rates drop and, at the same time, they may drool and exude a repulsive scent from their anal glands. This is often enough to cause predators that prefer live prey to lose interest. The catalepsy may last from a minute to several hours.

Whether feigning death is involuntary or a conscious decision on the part of the opossum is unknown, but it is one of the few mammals to employ this strategy, which is far more common among birds and reptiles. At other times, the opossum may defend itself quite vigorously, snarling and biting.

Perhaps no other animal has inspired such strange and varied myths as the opossum. Folklore ranges from the belief that the young cling to their mother's tail to the notion that the male copulates through the female's nostrils (probably inspired by the rather bizarre design of the opossum penis, which is forked). In fact, he couples in quite ordinary fashion with the female, whose vagina is similarly forked.

The folklore around opossum reproduction can hardly compete with the facts. The male courts a prospective mate by following her around and clicking his teeth continuously. When a receptive female eventually succumbs to his charms, she allows him to mount her from behind in typical mammalian fashion. During copulation, however, the pair flops over onto their right sides. If they remain upright or fall to the left, the female will not be inseminated. In addition to the forked genitals, the sperm are also twinned and they can only swim properly in pairs. If a pair of sperm is separated, each half swims in circles.

The development of the young is just as unusual. The embryos of most mammals grow a placenta, a pad rich in blood vessels, which becomes attached to the wall of the womb. Nutrients are transferred from the mother's blood, through the pad and the attached umbilical cord, to the embryo. With this rich food supply, the embryo grows to a fairly advanced state of development before leaving the mother's body. Once outside the womb, the transfer of nutrients to the young mammal continues via the mother's milk.

Without the benefit of a placenta, the young marsupials leave the womb much earlier in their development. They don't go far. In the case of the Virginia opossum, up to 22 tiny embryos, each no more than about 1.25 centimetres (½ inch) long, crawl from the birth canal just 12 days after fertilization. The entire brood can fit in a teaspoon. Their pink bodies are naked and their organs are barely formed. The tails and hind limbs are only stubs. Their forearms and front paws are well developed, however, and these are all they need to grasp their mother's fur, pulling themselves toward her pouch some 10 centimetres (4 inches) away.

But in what amounts to the world's cruelest game of musical chairs, when they arrive they find only 13 nipples. Each baby opossum that reaches a nipple clamps its mouth to it and begins to nurse. There is no sharing, as they will remain attached to the nipple for about two months. Numbers 14 and over are out of luck, and within hours their short lives are over. Sometimes, 13 or fewer opossums are born and everyone survives.

Those that are successful in claiming one of their mother's elongated nipples crawl about the pouch, exercising their limbs and developing all of the features that placental mammals develop inside the womb. Still hanging onto the nipples, they emerge from the pouch at about eight weeks, grasping her pelt with their paws. Shortly after this time, they adopt a more conventional nursing schedule.

For another 30 to 40 days, they cling to their mother with their prehensile tails, watching her forage and feed. As opossums age, their ability to hang by their tails gradually diminishes. They become too heavy to support the full weight of their growing bodies. Throughout their lives, though, they may still use their tails to drag loads of leaves or other nesting materials back to their dens, which may be an abandoned woodchuck burrow, a crevice in the rocks, a tree hollow, or an attic.

RIGHT: *Raccoons remain with their mother over their first winter, often traveling in bands of three or four. People seeing such a group assume it to be a family, but as with most mammals, the father has nothing to do with rearing his offspring and adults traveling together are likely to be females. Strange raccoons encountering one another usually greet each other with growls and other threats.*

Opossums are rather restless, solitary creatures, and never seem to occupy any one den for long. Their longest stays occur during prolonged cold spells or when raising their young. Some opossums seem to wander their whole lives, while others forage within a home range of from 20 to 200 hectares (50 to 500 acres). They are not strongly territorial, although males will fight viciously during the spring mating season. Quite common throughout their range, opossums seem to have overcome a lack of physical adaptations to cold through their willingness to eat almost anything. They live about seven years on average.

As unusual as the opossum is, its appearance is downright ordinary compared to the armadillo, surely one of the most bizarre-looking of mammals. Its body is roughly the size and shape of an old-fashioned toaster with a piglet's head at one end, and a lizardlike tail at the other. The animal is covered from head to tail in horny, brown scales of keratin that give the creature an almost reptilian appearance. One's first impression is of a tiny, armored dinosaur. The dome-shaped carapace that covers the body is unique among animals, composed of a series of bony plates joined by flexible bands. North America's only species, the nine-banded armadillo, has nine of these articulations.

ABOVE: *Before the arrival of Europeans, opossums were found no farther north than the state of North Virginia. Their range took a leap northward when a wave of human emigrants from the Appalachian mountains brought them to Washington state as game between 1890 and 1920. Many of these people considered the opossum a delicacy.*

LEFT: *The armadillo uses its keen nose to detect insects under as much as 15 centimetres (6 inches) of soil. It digs after the insects with strong, heavily clawed feet, and then slurps them up with its sticky tongue. In the course of a year, an adult armadillo may eat 90 kilograms (200 pounds) of bugs.*

The nine-banded armadillo is only the most recent of its kind to invade North America. Armadillos evolved about 50 million years ago in South America, and shortly after the land bridge joined the two continents, a few species worked their way north and eastward, some as far as the Ohio River Valley. One was the size of a black bear. But these early migrants were all wiped out during the major mammalian extinctions that began toward the end of the last ice age, about 10 000 years ago.

The nine-banded armadillo crossed the Rio Grande into Texas around 1850, and has been spreading northward ever since at a rate of 6 to 10 kilometres (about 4 to 6 miles) a year. They've recently been spotted as far north as Nebraska. They may be approaching the northern limit of their expansion, however, as they do not hibernate and have very little ability to accumulate fat. Armadillos have rather low body temperatures for a mammal and their shells provide much less insulation than fur.

Armadillos are found in semi-open, scrubby areas with loose soils that afford good digging. Rotting wood often harbors the ants and termites on which they most like to feed, so they prefer places with some trees.

Like other recent emigrants from South America, the armadillo owes its success to its catholic diet, but it also has the ability to deter most predators. Caught in the open, the nine-banded armadillo tucks its head under its tail and fans the articulations of its shell to enclose most of its body. Some of its South American cousins can actually tuck into an almost seamless lozenge. Smaller predators have difficulty getting their mouths around the armadillo's carapace, but the teeth of a cougar or a coyote can penetrate the shell, so armadillos prefer to run for cover if there is any nearby. They often escape into thorny underbrush, where their armor protects them and predators are reluctant to follow.

Armadillos are powerful diggers and if they are caught in the open, they will attempt to burrow to safety. Even if they can't bury themselves completely before being overtaken, they can wedge themselves into a partially dug hole with their powerful legs, leaving only the shell exposed. No armadillo actually escapes its enemies by tucking into a wheel and rolling downhill, as the old folk tale has it.

An armadillo on the run may also head for the nearest body of water. Although they hardly look like aquatic animals, they can ford bodies of water by one of two methods: Their shells are heavy enough that armadillos sink, so on coming to a stream, they often simply walk across the bottom, holding their breath for up to six minutes. Alternatively, they

swallow quantities of air, filling their intestines so that they become buoyant. Armadillos are surprisingly good swimmers.

Unfortunately, they're less adept at crossing pavement than they are streams. Roadways pose a hazard for all animals, but especially small, rather nearsighted animals such as the armadillo. They have a habit of arching their back and leaping straight up as high as 60 centimetres (2 feet) when startled. Although this reaction may in turn startle natural enemies, giving the shy armadillo a chance to escape, it has no such effect on an automobile passing over it at highway speeds. Many an armadillo that escapes the wheels of a car dies by crashing into the undercarriage. Today, more of them are probably killed by cars than by any natural predator.

Still, the armadillo is thriving. In the short 150 years since they entered the United States, their numbers have grown to somewhere between 30 and 50 million. In many areas they're considered garden pests, although they probably do far more good than harm to most gardens by eating considerable numbers of insects and other arthropods, which make up the bulk of their diet. The average adult armadillo eats 90 kilograms (200 pounds) of invertebrates a year. It has a sensitive nose and can smell beetles, larvae, or worms underneath the soil. It quickly digs a hole and then slurps them up with a long, thin tongue like an anteater's. It also eats amphibians, lizards, eggs, and some plants.

Armadillos mate in summer, usually in their burrows. Later, the female lines hers with leaves or grass in preparation for the birth of her young. As with many mammals, the armadillo delays implantation of the fertilized egg in the uterus wall until early winter. Recently, when a female that had been isolated in a laboratory gave birth, it was discovered that implantation of the egg can be delayed for at least two years.

Left: *In the arid southwest where trees are scarce, ringtails frequently den in rock crevices. Here, they are protected not only from the heat of midday, but many potential predators. At night, when they are active, they must contend mainly with owls and bobcats.*

Raccoons carry canine distemper and rabies, diseases that have increased with the urbanization of North America. One of the worst rabies outbreaks occurred in the U.S. mid-Atlantic region during the 1970s. Raccoons transplanted there for sport hunting introduced a strain of rabies to a population that had formerly been disease-free. People feeding raccoons also facilitate the spread of distemper by bringing the normally asocial raccoons in close and repeated contact with one another.

Females produce only one egg a year, but after fertilization and implantation it divides into four embryos of the same sex. Armadillos are the only animals that regularly give birth to genetically identical quadruplets. They are usually born underground, looking much like their parents in form except that their shells are pink and quite pliable. Over the first few days of their life, the shell darkens and grows harder, although it doesn't harden completely until the armadillo reaches full size.

They stay close to their mothers, learning how to forage during the spring and summer. With the onset of the summer mating season, they strike out on their own to begin their solitary adult lives, which may last 15 years.

Although the southern invaders are from unrelated mammalian families, they are more alike than different in several key ways. They all are relatively small, nocturnal animals of little commercial value. Raccoons are occasionally hunted for their pelts, and armadillos are of interest to medical researchers because they are one of the few species, besides human beings, to contract leprosy. While few appear on hunters' lists of preferred game, all have been hunted for food—particularly during the Depression, when many people were sustained by whatever game they could find.

For the most part, these species have benefited from human domination of the landscape. Urban development has eliminated many of their predators, and in the case of the raccoon, provided both food and shelter. The opossum and the armadillo have profited from the spread of monoculture agriculture, which has increased the numbers of insects available to them. At the same time, they seem to have escaped many of the most ecologically damaging aspects of human civilization. None requires large territories, and armadillos appear to be resistant to pesticide poisoning.

Ironically, we can measure the success of these species by the abundance of their dead, whose bodies litter our roadways. In one breath, we may profess our love of animals, and in the next, curses roll like sweat as we swerve to keep them from under our wheels, fence them from our gardens and, bathrobe-clad, collect the garbage from our lawns. But if we cannot make a place for these, the humblest and most accommodating of animals, what hope is there for any of our wildlife?

BELOW: *The Spanish word armadillo means "little armored one," and aptly describes the only mammals with true shells. The carapace of the nine-banded armadillo is composed of about 2500 bony plates covered in scales. Although the shell is not particularly strong, its smooth surface makes it difficult for larger predators to position their mouths for a firm bite.*

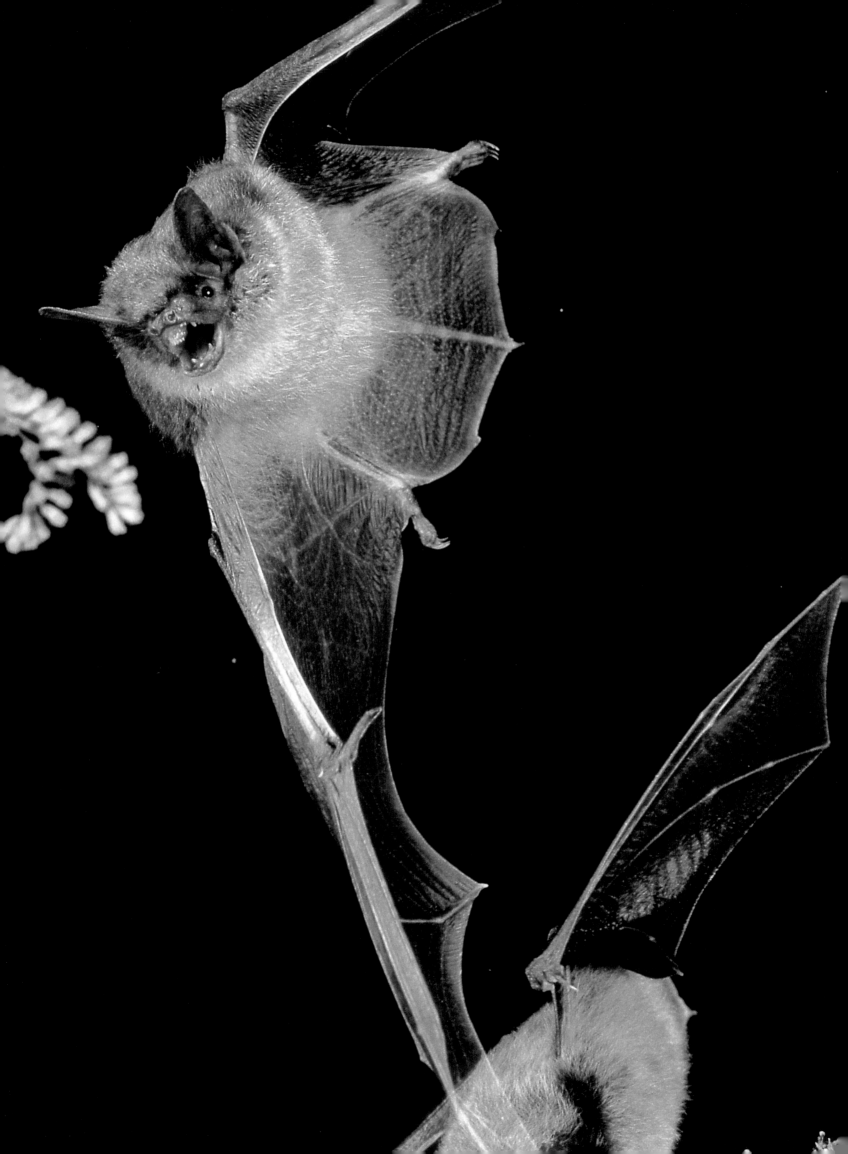

Air Superiority

Sixty million years ago, bats were the only mammals capable of sustained flight. They held the claim until 1783, when the first sheep flew. But the sheep cheated. It enlisted the aid of two other mammals, Joseph and Étienne Montgolfier, who set it aloft in the first successful flight of a hot-air balloon. A duck and a rooster accompanied the sheep on the trip.

Of course, birds had invented flight 80 million years before bats, and reptiles 20 million years before them. And all vertebrates are latecomers to the art of flying compared to the insects, which had taken wing some 200 million years earlier. But the mammals repeated one of nature's most successful experiments with the novelty born of ignorance. And indeed, nothing else flies quite like a bat.

Bats probably evolved from early shrewlike mammals that scampered over tree trunks and branches in pursuit of insects. One theory proposes that webs of skin stretching between the forelimbs and the rib cages of these insectivores may have been used to cup flying insects. Later, these flaps may have allowed the proto-bats to glide from branch to branch, eventually evolving into full, flapping wings.

ABOVE: *This big brown bat was photographed in a mine adit in southern Ontario. The sealing of abandoned mines has killed millions of bats, either by excluding them from these ideal hibernacula, or trapping many others inside. Today, bat conservationists advocate sealing mines with grates of a width that will bar people but admit bats.*

LEFT: *The bones supporting a bat's wings have evolved from the finger bones of the paws of its ground-dwelling ancestor. As few bat skeletons are fossilized, no one knows exactly what this proto-bat was like. It may have resembled today's shrews, with membranes that stretched between the ankles of its hind and forepaws, much like those of today's flying squirrels.*

Townsend's big-eared bat was named for J.K. Townsend, an ornithologist and author whose name has been given to at least a dozen species of mammals. His ears were of quite ordinary size. The bat uses its huge ears to detect the echoes of its lower-pitched calls. These sounds of longer wavelength are probably used to detect whole swarms of flying insects at a distance. Higher-pitched squeaks are used to zero in on individual insects.

Bats do not fly as efficiently as birds. They have never developed two of the birds' greatest assets: feathers and hollow bones. Nor do bat skeletons have the deep keels and large attached breast muscles of the birds. They are not particularly good at gliding or soaring, which all of the long-distance avian fliers use to their advantage. Still, it beats walking. While a bat flying from one point to another consumes calories at many times the rate of a mouse scampering the same distance, it gets there so much faster that in the end, it uses far less energy.

What the bats lack in endurance, they make up for with maneuverability—both in the air and when roosting. Bats can chase individual insects through forest canopies and can hover, dip, and swoop in pursuit of their prey with an ease unmatched by birds. The absence of a bony keel and their broad flat bodies enables them to squeeze into narrow crevices for day roosting and hibernation.

Almost all of the bats of North America eat only insects. In the temperate zone, where fruits and flowers are available only seasonally, there simply isn't enough food to build the fat reserves necessary to survive northern winters. Some pollen- and fruit-eating bats do inhabit the southern United States and Mexico, while fishing and vampire bats live only in the tropics, but the range of insectivores such as the little brown bat extends as far north as central Alaska.

The bats are a highly successful order of mammals. One in four mammalian species is a bat; only the rodents have more. Although only 40 of the world's 1000 species live in North America, in the warmer parts of the United States and Mexico they are among the most numerous of vertebrates. There is a cave in Texas where 20 million Mexican free-tailed bats roost each night.

Their success stems from their flying abilities, in particular their ability to fly at night. By feeding after dark, bats have almost sole access to an enormous food source: nocturnal flying insects. During the day, birds are the chief hunters of flying insects. But hunting after dark is hazardous for a low-flying bird, and trying to see something as small as an insect after twilight is just about impossible. Before bats, the insects had the night sky to themselves.

But the bats had already developed a trick that allowed them to overcome both these problems before they ever got off the ground: sonar. Like some shrews today, the proto-bat may have used echolocation to find its insect prey. With active sonar, or echolocation, modern bats can navigate

ABOVE: *Like many North American bats, little brown bats are true hibernators. Their body temperatures drop to within a few degrees of air temperature and their hearts may slow to only three or four beats per minute. A special tendon locks their needle-sharp claws into the closed position, so clinging to irregularities in the ceiling of a cave requires no effort over the long winter.*

obstacles and locate insects as small as a mosquito in complete darkness.

The principles are the same for bats as they are for dolphins, submarines and some shrews: The bat emits a squeak using its vocal cords and then listens for the returning echo. As the speed of sound in air is relatively constant, the time taken for the echo to return indicates how far away the obstacle is. The volume and other qualities of the echo reveal much to the bat about the size and even the shape of the target. By emitting repeated squeaks, a bat builds up a sonic image of its surroundings.

The word "squeak" is somewhat misleading. The lowest sounds made by most bats for echolocation are of too high a frequency for most people to hear. It's a good thing. The call of the little brown bat, one of the most common mammals in the world, is roughly the same volume as the alarm on a smoke detector. If it were not well beyond the range of human hearing, the resulting noise might drown the urban din of car alarms.

The spotted bat, on the other hand, squeaks at 6000 to 16 000 cycles per second—a sound range audible to anyone with normal hearing. Many of the sounds are so brief that they might more accurately be called clicks. The higher the squeak, the higher the resolution of the sound pulse, and the smaller the targets it can detect. Bats using lower squeaks can't locate quite such small insects, but they can detect them from much farther away. This makes sense as a big beetle several seconds' flight away might be enough of a meal to justify a detour, while a gnat probably doesn't have enough calories to be worth the bother.

Bats take such a toll on insect populations that the insects have fought back with adaptations of their own. Some moths and lacewings can hear bat calls and fly away from them. If the bat is closer, they drop to the ground or take evasive maneuvers. There are tiger moths that can respond with their own clicks by rubbing together ridges in their exoskeletons. This may work to the moth's advantage in one of two ways. The moths may taste foul and after a few bad experiences, the bat comes to associate the click with the bad taste and breaks off the attack. Or, the resulting sounds may actually jam the bat's sonar, making it impossible for the bat to hear the echo and find the target.

The insects have an edge in this sonic war: They can hear the bat from at least twice as far away as the bat can hear the insect. That's because the echo must travel to the target and back for the bat to detect its prey; the insect has only to listen to the outgoing ping.

Recent experiments may have determined the limits of resolution in bat sonar. Remarkably, the big brown bat can distinguish objects that are only 0.3 millimetres (0.0125 inches) apart. This is about three times the resolution of the best artificial sonar, and no one is quite sure how the bat can process signals so precisely. What's even more amazing is the speed with which bats can process and act upon what they hear.

One experimenter put a bat in a box about the same dimensions as a three-drawer filing cabinet lying on its side. The bat had a wingspan of 25 centimetres (10 inches), only 5 centimetres (2 inches) less than the depth of the container. It was able to fly for several minutes without once touching the sides of the box.

Many bats have enormous ears relative to their bodies to cup and pinpoint their sonar echoes. These add drag, but bats are designed more for low-speed maneuverability than for speed or long-distance flying. Still, bats are quite capable of fast and long flights. In a single night, they may travel as far as 95 kilometres (60 miles) from their roosts to feed. Swarms traveling up to 105 kilometres (65 miles) an hour have been tracked by radar. Special acoustic detectors sent aloft on helium balloons have recorded the calls of bats hunting at altitudes of 3000 metres (9800 feet).

In the tropical rainforest, bats are vital to the dispersal of seeds, but in temperate climates their main ecological impact is the consumption of huge numbers of insects. Bats are voracious, eating up to half their own weight in prey each night. For an average-sized bat, that may mean 600 insects an hour. Insect-eating bats have mouths full of sharp, conical teeth resembling those of the shrews. They use them to crunch through the shells of large flying insects.

With the return of daylight, bats must find a safe place to rest and hide from predators. Some roost in caves or mines, where they may hang from the roof or walls with thousands of other bats. Others take to rock crevices, tree hollows, attics, or the spaces between roofing shingles— sometimes in large groups and other times alone. For hanging from ceilings, bats have special tendons that lock their needle-sharp claws to the tiniest knob or cranny. Their ability to wedge themselves into narrow spaces is remarkable.

In the early 1980s, renovations to the Constitution Bridge in downtown Austin, Texas inadvertently created a perfect day-roosting habitat for Mexican free-tailed bats. Today, a million and a half bats squeeze into the spaces between the supporting beams, spending their days in concentrations of up to 5000 bats per square metre (500 per square foot). Their nightly departure has become something of a tourist attraction. Many biologists believe that the readiness of bats to adopt these artificial spaces is an indication of just how scarce prime natural roosts have become.

Insect-eating bats are among the smallest of mammals. For any small warm-blooded animal, temperature regulation is a problem—particularly during our northern winters. It's especially difficult for bats because their wings, which account for a large part of their body's surface, cannot be covered in insulating fur if they are to function as air foils. To replace the heat lost through these large exposed areas, bats have rapid

RIGHT: *Mexican free-tailed bats exit Carlsbad caverns in New Mexico at dusk to feed on insects. They stream from the cave, peaking at 5000 to 10 000 bats per minute. In the 1930s, 8 or 9 million bats roosted in these caves. Today, there are perhaps half a million, their numbers devastated by insecticides. Free-tailed bats are still among the most numerous of mammals in North America.*

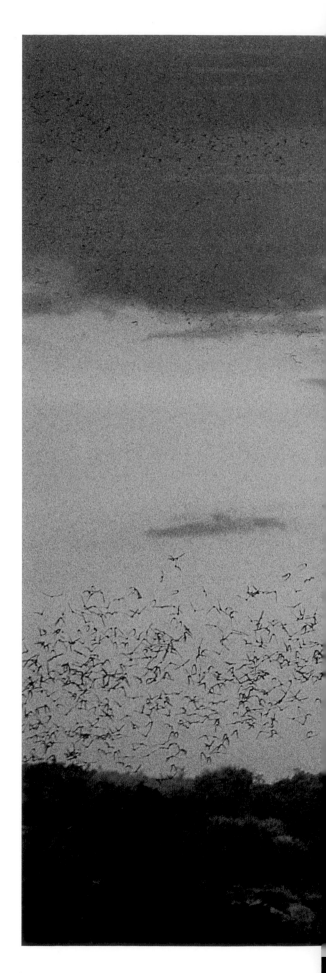

metabolisms. The heart of a bat in flight buzzes rather than beats, at 20 times a second—about the same rate as a hummingbird's.

A rapid metabolism must be fueled by constant eating and during the winter, flying insects vanish from most of North America. Bats in northern regions respond by hibernating, but first must find a suitable roost in which to pass the winter. Ideally, the roost should stay at a stable temperature. It cannot get too cold or the bat will freeze to death. Nor can the roost be too warm. If the bat's body temperature does not drop below 10°C (50°F), its metabolic rate will remain too high, the bat will burn all of its fat stores before spring, and it will starve. The ideal for many temperate species appears to be around freezing. At these temperatures, a bat's heart may beat as slowly as three or four times a minute. It may go up to an hour between breaths.

Within a cave or building, bats can adjust their body temperatures by choosing where to hang. Usually, it is warmest near the ceiling and cooler toward the floor. Bats may form clusters of hundreds or roost separately to further regulate their body temperatures. Every three weeks or so, they rouse themselves from their torpor to drink and stretch.

If their body temperature approaches freezing despite these efforts, some bats will actually seek a new roost in the dead of winter, but such a move is costly, burning large amounts of body fat. It may mean the difference between surviving the winter and starving to death before the spring insect hatches. For this reason, it's vital not to disturb any bat you see roosting in winter.

Bats also need day roosts during the summer to nurse their young and to rest. Scientists in Arizona studied a colony of lesser long-nosed bats, a species that feeds on nectar, pollen, and fruit. They entered the cave, set up infrared lights, and aimed remotely operated cameras at the bats hanging from the cave ceiling. Bat mothers have been known to cooperate by nursing and babysitting one another's infants during foraging absences, but that was not the case with the lesser long-nosed bats.

The scientists did find that mothers, who returned an average of three times each night to check on their young, had an unerring ability to find their own pups among the 11 000 bats roosting in the cave. They nursed their young twice during the night, and neither mother nor infant appeared to spend any time sleeping. Bats seldom give birth to more than one pup at a time. Considering that the pup may weigh 40 percent as much as its mother, it's obvious why. The young must be well developed

LEFT: *The seminole bat is solitary, by day hanging in clumps of Spanish moss wherever the plant is found in the southeastern United States. At sundown, they leave their roosts to hunt insects flying around the crowns of trees or over creeks and ponds. They sometimes alight to catch crawling insects. The females often give birth to twins.*

ABOVE: *During World War II, the American military devised a plan to strap tiny incendiary bombs to thousands of Mexican free-tailed bats and release them from bombers over Japanese cities. The bats, it was reasoned, would roost under the eaves of the wooden buildings, where the bombs would drop off and ignite. Tests were successful, but the plan was eclipsed by the Manhattan Project.*

RIGHT: *The eastern red bat is a solitary bat that has up to four pups in a litter. In summer, they are found as far north as south-central Manitoba, but in winter most migrate to the southern United States or Mexico. Even in these warm places, the red bat spends much of the winter hanging from vegetation and dozing in a state of torpor.*

from birth to hang from cave walls or ceilings and survive their mothers' foraging absences. To compensate, bats have long reproductive lives. Some species live 20 or 30 years and may begin bearing young as early as their first year.

At best, bats' relationship to human beings is uneasy. We rarely get a good look at them, but what glimpses we have of their furry bodies, leathern wings, and sometimes bizarre facial appendages supply plenty for the imagination to work on. That our European ancestors seem to have drawn heavily on the bat's body for their depictions of gargoyles and demons has not helped. Angels' wings are feathered, but Satan has the naked wings of a bat.

With the writing of *Dracula* in 1897, Bram Stoker put the last nails in the coffin of the bat's public image.

There is only one real reason for people to fear bats: They are one of the main carriers of rabies. Recently in North America, there have been a few cases in which people have awakened to find bats in their rooms and shooed them out of a window. Months later, the people developed rabies and died. Doctors could only surmise that the bats scratched or bit the unfortunate victims as they slept, although the wounds were so slight

they were never detected. By the time a rabies victim shows symptoms of the disease, there is almost no hope of recovery.

In fact, the percentage of wild bats carrying rabies is rather small, and fewer than 10 people in North America die from all sources of the disease each year. The trouble is that as we destroy more of the bat's natural roosting places, they become more and more dependent upon our buildings, mines, and bridges for their survival. Consequently, people and bats are coming into more frequent contact.

Currently, there are eight species of bats listed as endangered in North America, among them the Indiana bat, whose numbers have declined by half since 1983. Bats are threatened by deforestation, the razing of old buildings, the sealing of abandoned mines, recreational spelunking, and the spraying of agricultural pesticides, which accumulate in their bodies over years of eating insects. This last danger points out a particular irony, as the conservation of bats is probably the single cheapest and most effective method of pest control available. Some farmers who have erected bat roosting boxes near their fields have been able to eliminate the use of pesticides.

There are some signs that the public's attitude toward bats is improving. In 1987, when an old building housing a colony of about 1000 Brazilian free-tailed bats on the campus of the University of Florida burned down, they took up residence in a newly built sports stadium. The sight of bats swooping over the bleachers as they hunted insects during night games was unnerving to many of the fans, but the Brazilian free-tailed bat is protected by law in Florida.

Rather than exterminate the bats, the university constructed a 6-metre-square (20 feet by 20 feet) bat house some distance away from the stadium. They trapped a few dozen of the stadium's bats and introduced them to the house, but as soon as night fell, they fled. For five years, the $20,000 house stood empty. Early in 1993, however, some 18 male bats roosted there briefly of their own accord and then left. A little later, about 300 females visited. They also left. But a few months after that, the bats arrived to stay. By May 1998, 70 000 bats had taken permanent residence in the bat house, which is capable of accommodating three times that number. A second bat house with room for half a million bats is currently under construction. Although bats still frequent the stadium's night games, the smell of guano—and any mosquitos—have left the stadium.

Bats are one group of animals whose decline could be reversed with little more effort than a change in attitude.

Those That Wait

There is nothing like the patience of a reptile. While birds and mammals burn their foolish furnaces, the masters of the lunge wait with only the pilot light flickering.

Because reptiles do not rely upon metabolic processes to heat their bodies, their caloric requirements are much lower than those of the warm-blooded animals. As a result, they can wait a very long time between meals. In the tropics, some of the large constricting snakes need to feed only once a year. Here in North America, a rattlesnake probably consumes only three times its body weight annually, while a mole eats three times its body weight in prey every week.

Freed of the need to hunt or forage constantly, many reptiles live long and sedentary lives. A 50-year-old snapping turtle may lie at the bottom of a pond for weeks in exactly the same spot, moving only its neck to lunge at a passing fish or to bring its nostrils to the surface for a breath every hour or so. In hibernation, they may pass the entire winter without drawing a single breath at the bottom of some ice-covered pond. A few turtles are able to absorb some oxygen through their skin, prolonging the time between breaths.

ABOVE: *During the day, the young alligator snapping turtle lures fish to within striking distance by opening its mouth and wiggling a tiny, reddish, wormlike appendage to its tongue. At night, when this visual lure is ineffective, the snapping turtle searches more actively for aquatic weeds, mollusks, and fish.*

LEFT: *The American alligator is arguably the continent's most formidable ambush predator. With its ability to approach prey undetected, large size, and high lunge speed, it is formidably equipped to catch and subdue large prey. With their comparatively low metabolic demands, these reptiles can go for weeks between meals.*

Patience and the ability to lie motionless for long periods make reptiles master ambush predators. Turtles, many lizards and snakes, and alligators spend much of their time lying in quiet wait for their prey. When an animal strays within range, the hunter lunges with a single, swift movement. While most reptiles can move quickly as long as they are warm, few of them can sustain any kind of vigorous activity for more than a few tens of seconds. Their respiratory and circulatory systems are just not as efficient as those of birds and mammals. In one exception, male snapping turtles thrash about in the water for tens of minutes during the spring battles for territory.

Reptiles depend entirely on external heat sources to keep their bodies in their ideal temperature range—which is usually about the same as the body temperature of a mammal or bird of equivalent size. They regulate their body temperatures by moving in and out of direct sunlight, burrowing, climbing bushes, or lying on rocks or soil warmed by the sun. Turtles sometimes bask underwater. At night, many of these methods of thermoregulation are unavailable in all but the southern states. By dawn, most reptiles are chilled and vulnerable, unable to move quickly until they've had a chance to bask or drape themselves over a warm rock.

Lasting through the night is a minor problem compared to surviving a northern winter. Reptiles are completely inactive during the winter throughout most of North America, and the number of species drops rapidly with rising elevation or latitude. Sagebrush lizards live as high as 1800 metres (6000 feet) in the Sierra Nevadas of California, but spend half of every year in hibernation. Garter snakes have the most northerly range of Canada's reptiles, living just below 65°N, about the latitude of central Yukon. Their period of activity is even briefer, perhaps three months. Many turtles survive in northern latitudes by passing the winter

RIGHT: *The chuckwalla is a fairly unusual lizard in that it eats only plants. Confronted by almost any animal larger than itself, the chuckwalla retreats to a crevice in the rocks. By swallowing air, it can inflate its body to three times its normal size, making it nearly impossible to dislodge from its hiding place.*

in lakes and ponds that don't freeze to the bottom. Despite these adaptations, few reptiles live north of the 49th parallel.

By comparison, the southern United States has a rich reptile fauna. The scaly skins of snakes and lizards are practically impermeable to moisture and have few glands. Their excretory systems are also efficient at conserving water, so many species thrive in the deserts of the southwest where few other vertebrates can live.

Of the three main reptile groups—lizards, snakes, and turtles— lizards are the most numerous, with some 3000 species worldwide. Of these, about 115 live in North America, but only 5 are found north of the Canadian border. The first reptiles to evolve, about 300 million years ago, followed the basic lizard body plan.

Lizards that hunt during the day see well, while those that hunt underground or at night usually have poor vision. Among nocturnal species, smell is the primary sense. At a distance, lizards and snakes take in odors through their nostrils, much as we do. But for smells from nearby sources, they also taste the air with their forked tongues, picking up scent particles, then drawing in the tongue to wipe them on the Jacobson's organ on the roof of the mouth. Nerves lead from this organ directly to

ABOVE: *There are at least five subspecies of short-horned lizards in North America, all living in dry areas throughout the western half of the United States and southern Canada. Like many of the horned lizards, short-horned lizards eat mainly ants and a few other insects. They are active during the heat of the day, and at night they bury themselves in sand or soil.*

LEFT: *Bullsnakes inhabit a wide band from southern Alberta to Mexico, and often enter the burrows of prairie dogs and ground squirrels in search of prey. Although not venomous, they may react aggressively to people, hissing, recoiling, and occasionally striking. They usually hunt by day, but in the hotter parts of their range they may become nocturnal during the summer months.*

ABOVE: *Up to 75 000 red-sided garter snakes slither from as far as 20 kilometres (12 1/2 miles) away to pass the winter near the town of Narcisse in southern Manitoba. They congregate by the hundreds in sinkholes in the limestone bedrock. Although the bodies of many of these snakes will drop below freezing over the winter, most of them will emerge none the worse in the spring. This mass gathering is unique among reptiles.*

from their burrows in the morning and late afternoon to hunt, avoiding the heat of midday. Even when their bodies are at their optimum temperature, they move sluggishly, with a confidence reserved for venomous animals.

Their system of venom delivery is more like a shrew's than a snake's: Glands open into grooves leading from the bases of the teeth along the bottom jaw. When the Gila monster bites, the poison is drawn up the groove to the tip of the tooth and the lizard works it into the wound with a chewing motion. For such a system to work, the Gila monster must be immune to its own poison. They eat mainly the nestlings of hares and rodents, birds, birds' eggs, and other lizards. Seldom is their bite fatal to human beings and, unlike many poisonous snakes, a person would have to really aggravate or step on a Gila monster to provoke it to bite.

As with all reptiles, their eggs are fertilized internally and laid on dry land—in this case, a shallow depression the female digs in the ground. Courtship coincides with the midsummer rains, and seems to consist largely of male and female flicking their tongues at each other. The 3 to 12 eggs overwinter and hatch the following May. The mother offers no care to her eggs or young, which is typical of reptile parents.

The five-lined skink lives throughout the eastern United States and is Ontario's only lizard. In sharp contrast to the Gila monster, skinks are thin and agile, and dart quickly away from predators. There are quite a few color variations among the adults, but all display the nominal five lines running down their backs. The most startling aspect of their coloring, however, is the electric-blue tail of the juveniles.

The purpose of the bright color might be to decoy predators. Like many lizards, the five-lined skink can shed its tail, which continues wriggling after it has been detached. This may distract the predator long enough for the rest of the lizard to escape. At a particular spot in the tail, the bone and the muscles are pinched and thin, allowing it to break off with very little force. The lizards hardly bleed from their stumps and seem to suffer no permanent harm. When the tail regenerates, it has no vertebrae in it from the breaking point on and is usually somewhat stumpier than the original tail. It can regenerate as many times as it's broken off.

The problem with explaining the blue tail as an added distraction is that if this strategy is effective for the young lizard, there seems no reason why it shouldn't be of just as much benefit to an adult skink. And yet, the blue fades within a few months of the lizard's hatching.

Five-lined skinks inhabit mainly woodlands, hunting over the moist forest floor for insects, spiders, and worms—almost any invertebrate small enough to swallow. They, in turn, are eaten by just about any carnivore or raptor large enough to swallow them. In the northern part of their range, they spend the winter wedged into rock crevices or squeezed beneath logs, their bodies nearly frozen, in a state of profound hibernation.

The female five-lined skink remains curled around her clutch of up to 18 eggs, which she lays under an overhanging log or stone or buried in the leaf litter. She cares for the eggs by turning them occasionally, but within a few hours of their hatching, she loses all interest in her brood and abandons it.

The glass lizard is another lizard that wraps itself around its eggs— although no one has ever seen one actually defending its brood. There are four species in North America and one of them, the eastern glass lizard, grows to 105 centimetres (41 inches). Their name comes from their readiness to shed their tails if handled.

Glass lizards look like snakes in that they have no legs, but on closer examination other features that identify them as lizards are visible. These include eyelids, external ear openings, and nonexpandable jaws. Many people still call them glass snakes. The lack of legs helps them to burrow in loose soil—sometimes as deep as 30 centimetres (one foot)—in search of earthworms, or among leaf litter where they hunt for crickets, spiders, grasshoppers, and other small invertebrates.

The distinctive feature of the turtles is, of course, the shell, which is strong enough in most species to protect them from the jaws of predators many times their size. When threatened, the turtles draw in their head, legs, and tail and a predator is faced with a tough nut to crack. Even if an animal has jaws strong enough to crush the shell, getting them around a large turtle is just about impossible. Most predators eventually give up, at worst flipping the turtle onto its back before leaving. Once again, reptilian patience pays off. The turtle simply waits until its persecutor leaves before righting itself. It's a simple defense, but one that has served the turtle well. In their 250-million-year history, they have watched the rise and fall of the dinosaur with almost no change to their own strategy or design.

It's a good thing for the turtles that their shells are so effective because they're also so heavy that these reptiles are incapable of outrunning even the slowest predator. Turtles, however, are well adapted to life underwater with long breathhold times and broad feet that enable them to swim rapidly. Most turtles pass the winter underwater, and in the northern part of their ranges many remain active foraging even in ponds or lakes that are completely iced over. They eat mainly fish, tadpoles, snails, worms, and other invertebrates, but many species also feed on algae and aquatic plants. When these foods are hard to come by, the turtle may park itself on the bottom or burrow into the mud, withdraw its limbs, and remain in a torporous state until the water warms up. But even on a winter day, if the temperature rises above about 10°C (50°F), turtles may pull themselves onto logs or rocks rising above the surface to bask. They can push themselves into the water in a second if threatened.

The shell of the alligator snapping turtle offers little protection for its limbs. The plastron, or lower shell, is small, and the snapper cannot draw its limbs completely inside the shell the way other turtles can. Instead, it relies on its beaklike mouth and powerful jaws for protection. The alligator snapping turtle is the largest of all fresh-water turtles, reaching weights of up to 27 kilograms (60 pounds) and its bite should be respected by people wanting to keep their fingers.

The tortoises, which never enter water, rely entirely upon their high,

domed shells for protection. Because they are all vegetarians, they do not have to move quickly in pursuit of prey.

In territorial fights, male gopher tortoises of the southern United States attempt to use their shells against each other. Combatants broadside one another like drivers in a demolition derby, each seeking to overturn his opponent, thereby removing him as a mating competitor, at least temporarily.

As if its shell weren't enough protection, the gopher tortoise uses its strong and heavily clawed forelegs to dig a burrow that may be 9 metres (30 feet) in length. For its efforts, the gopher tortoise often ends up having to share its quarters with foxes, raccoons, burrowing owls, rattlesnakes, skinks, and at least 14 species of insect that live nowhere else.

As if the reptiles had grown bored after some 50 million years of success with their plodding, armored tanks, about 150 million years ago they tried pushing the lizard body in the opposite direction, an experiment that resulted in nature's most radical departure from the tetrapod body plan.

Snakes are the embodiment of grace: effect without effort. How the undulations of their limbless bodies results in forward movement appears as mysterious as it does beautiful. In fact, snakes move by four different methods, but the one employed by all snakes is serpentine motion. In this method, the snake bends its body into a series of lateral loops by contracting the muscles on the inside of each loop. These areas of contraction pass in a wave from head to tail, as do the loops. Those parts of the loop that are nearly perpendicular to the snake's direction of travel push against irregularities in the ground. This forces the snake's body forward. Snakes also use serpentine locomotion when they swim.

None of the methods of snake locomotion is very fast and few can even

LEFT: *Few reptiles are as familiar as the painted turtle, which has the greatest range of any North American turtle—a wedge with its point at the Strait of Georgia and fanning to almost the entire eastern seaboard. The painted turtle often basks by the dozen on logs or rocks. On one occasion, painted turtles climbed onto the back of a nesting loon, which ignored its house guests.*

ABOVE: *The tracks of this sidewinder (a species of rattlesnake) trace its characteristic lateral motion. In addition to being the only kind of snake locomotion that is effective on sand, sidewinding has another advantage: As the tracks show, only two points on the snake's body are actually in contact with the ground at any one time, which helps to keep it cool in the deserts of the southwest where it lives.*

keep up to a strolling person, although the serpentine motion makes them seem to move much faster than they really do. Nor is slithering particularly efficient. It takes a lot of lateral body movement to achieve any forward progress. So what could be the advantage of limbless locomotion?

The main benefits appear to be near silence and the ability to pursue prey into tunnels hardly greater in diameter than the snake itself. Snakes also climb trees and bushes well. And a coiled snake has no regular shape or outline for the eyes of predator or prey to key on, so they're hard to spot even when lying in plain sight.

Snake sensory systems are similar to those of lizards except that snakes have no ears. They are deaf to airborne sounds. They are, however, extremely sensitive to vibrations conducted through the ground and can detect the footsteps of approaching animals. Their vision is also generally poor. As with the lizards, smell—or more accurately, their ability to taste the air with their tongues—is probably their most important sense for detecting prey at medium distances.

But as a rattlesnake follows the scent to its source, a sixth sense, the ability to detect the body heat of its prey, comes into play. The organs that allow it to do this are two pits—one between the eye and nostril on each side of its face. These pits are so sensitive that they can detect a contrast in temperature of as little as .003°C (.005°F). Heat detection works best at night, when the temperature contrast between a warm-blooded animal and its surroundings is greatest. The bottom of each pit is lined with special nerve cells that react to radiant heat. The pits themselves act like the body tube of a telescope, screening out stray radiation and allowing only heat sources directly in front of the snake to "shine" onto the sensory cells. In effect, when the heat source appears brightest to the snake, its head is aligned to strike.

A rattlesnake's fangs are hollow and located at the front of the mouth. When its jaws are closed, the fangs point backward and are sheathed by a fold of skin. As the snake strikes, its jaws open very wide (about 120 degrees) and the fangs swing forward to point at the target. The force of the strike stabs the fangs into the target's flesh. At the same time, muscles squeeze venom from the glands at the fangs' base through a hole at the tip, injecting the prey. These fangs are replaced about every three months, with the new ones growing in next to the old. The old ones are eventually lost when they become stuck in an animal's body.

The entire strike, from beginning to end of the recoil, takes about a quarter of a second. Within 30 seconds, an animal the size of a kangaroo rat will begin to falter from the effects of the venom. Most rodents run as soon as they are bitten and may travel up to a hundred metres (about 300 feet) from the site of the attack before collapsing.

Immediately, the rattlesnake begins flicking its tongue and sweeping its head from side to side in an attempt to detect the rat's scent. The tongue and Jacobson's organ soon pick up the trail left by the dying rat, which the snake follows. Once it reaches the rat, the snake prods it a few times with its snout to make sure that it is unconscious, then locates the head by smelling for odors from the rat's mouth, and begins to swallow its prey whole and quite possibly still alive.

Snakes have the unique ability to unhinge their jaws so that they can swallow prey greater than the diameter of their bodies. The lower jawbone is not a complete horseshoe, as in other vertebrates. Instead, the curved part of the U is replaced by ligaments that are highly elastic. Rattlesnakes regularly eat mammals as large as cottontail rabbits. The snake's backward-pointing teeth, while useless for chewing, help it to inch its jaws over the body.

The resulting distension of the head makes breathing difficult but the glottis, where the mouth opens into the trachea, is held open by hoops of cartilage erected by muscles during swallowing. Gradually, by squirming from side to side, ratcheting the teeth on one side of its mouth forward and then on the other, the snake's mouth gradually envelops the prey. A flood of saliva from glands in the mouth helps to lubricate the whole process. Swallowing a small rat may take a few minutes, a rabbit up to two hours, depending on the relative sizes of snake and prey. When it's done, yawning and stretching help to realign the jaws.

Digestion is already underway before the rat is completely swallowed because much of the rattler's venom is made up of digestive enzymes. With smaller prey, the rattlesnake immediately resumes the hunt for its next meal. If it has eaten a large animal, the snake may have to lie quietly and digest its meal—sometimes for up to a week. During this time, the closer it can keep its body temperature to its optimum of about 30°C (85°F), the more quickly digestion proceeds.

The strike-and-release strategy of the rattlesnake is efficient. It minimizes struggle with the prey and the commensurate danger of the snake being wounded by teeth or claws. Although producing venom is metabolically costly, it takes less energy than a prolonged struggle. And poisoning allows snakes to kill much larger prey than they could subdue by constriction alone.

But venom has another benefit: It poses a lethal threat to animals much larger than itself, and predators that would readily eat a non-venomous snake its size are hesitant to attack a rattler. The rattlesnake advertises its threat not by the distinctive markings of many venomous animals, but with an audio warning.

The organ for which the rattlesnake is named is a series of interlocking lobes made of keratin (the same substance our fingernails are made

RIGHT: *The eastern box turtle is highly terrestrial, living in lowland forests and meadows where it hunts for slugs, earthworms, insects, berries, and mushrooms. It is also extremely long-lived, and individuals kept as pets by a succession of human owners have reached over 130 years of age. This turtle is quite sedentary, and may spend its whole life in half an acre (one fifth of a hectare).*

of) at the tip of the tail. The snake is born with only one lobe, but each time it sheds its skin (up to several times a year, depending on its age) it leaves another lobe, interlocking with the lobe from the previous shedding. Older lobes are usually broken off over time, so counting lobes is not a reliable way to age a rattlesnake. The fit between lobes is imperfect, and they rattle against each other when the tail is shaken.

The snake shakes its rattle when it feels threatened. In this way, it may warn off potential predators or even herbivores that could accidentally trample it, without having to strike. Some biologists dispute the rattle's effectiveness as a warning, arguing that many hoofed animals are just as likely to try to stomp a rattler to death as avoid one. Certainly, the warning frequently backfires in confrontations with human beings, many of whom feel compelled to beat a rattler with a stick or shoot off its head in the name of public safety.

One predator on whom the warning is wasted is the kingsnake which, like all snakes, is deaf. Kingsnakes hunt and kill other snakes, including rattlers, and are immune to their venom. Their instinct to swallow other snakes is so strong that the young begin trying to eat one another within days of emerging from their eggs. Kingsnakes kill other snakes by biting

ABOVE: *Two male desert tortoises charge one another during a courtship dispute. Each tortoise attempts to overturn his opponent by hooking his scute—the bony projection below the neck—underneath his rival's carapace and pushing. Tortoises differ from turtles in that they carry their shells well off the ground when they crawl.*

LEFT: *The regal horned lizard can be distinguished from other horned lizards by the arrangement of spikes on its head. In the regal horned lizard, the bases of the spikes touch, forming a continuous fringe. Like a number of horned lizards, the regal will sometimes squirt blood from the corner of its eyes if threatened. How, or whether, this is an effective defense is not known.*

them on the head or neck and then suffocating them by constriction. They swallow their prey head first. Kingsnakes also eat small rodents, birds, their eggs, and amphibians. There are several species in North America, many of which grow to over 2 metres ($6\frac{1}{2}$ feet) in length.

The garter snakes neither poison nor strangle their prey, but swallow it alive. They eat amphibians, fish, and invertebrates. In the eastern part of their range, many are known to eat voles, but any larger mammal probably puts up too much of a fight for the garter snake to swallow it alive. A few giants up to 125 centimetres (50 inches) have been found, but most are under 60 centimetres (24 inches).

Garter snakes are the most widely distributed snakes in North America, living as far north as the Arctic Circle. At these latitudes, they spend more than two-thirds of the year in hibernation, often congregating in sinkholes or the burrows of mammals to pass the winter. Farther south, near the town of Narcisse, Manitoba, as many as 75 000 red-sided garter snakes slither up to 20 kilometres (over $12\frac{1}{2}$ miles) to hibernate in natural cavities in the limestone bedrock, many returning to the dens in which they hibernated the previous year. Curiously, none of these returning snakes are the young of that year. No one knows where the snakes spend their first winter.

In addition to being highly resistant to cold for a reptile, garter snakes live in a variety of habitats. In the eastern and southern part of their range, they tend to be more aquatic, hunting crayfish and frogs. Like most snakes, garter snakes are strong swimmers and have been seen up to 100 metres (330 feet) from shore, occasionally in salt water. In the west, they prefer damp meadows, back yards, and any kind of semi-open area where they can raise their body temperatures by sunning themselves.

RIGHT: *Also called the water moccasin, the cottonmouth is the only pit viper (a group including the copperhead and rattlers) that is primarily an aquatic hunter, though it also eats a variety of prey on land. Cottonmouths can be distinguished from less-dangerous water snakes because they inflate their lungs when swimming. Much of their body floats, with the head carried well above the surface.*

The garter snake's most effective defense—at least against people—is the foul-smelling musk the snakes exude from their cloaca. They attempt to smear it over their captors with their coils. Birds of prey, raccoons, skunks, weasels, mink, and sometimes chipmunks will all hunt garter snakes. Like many small animals that are relatively defenseless, garter snakes are prolific. They seem to rely on chance encounters to bring male and female together in spring for a fairly elaborate courtship, in which the male entwines himself with the female, rubbing her chin.

Snakes copulate by aligning their cloacas, following which the male extends one of two penislike organs called hemipenes into the female's cloaca. Why snakes need two hemipenes no one is sure, but in many species each hemipene in turn has a bifurcated head, giving the snake a surfeit of copulatory organs.

The female has her young in late summer or early fall. While most snakes lay a clutch of eggs and abandon them, the female garter snake bears anywhere from 6 to 50 live young, which look like miniatures of their parents.

Probably the best-known, and by far the largest, reptile in North America is the alligator. The record specimen, caught in 1890 in Louisiana, was almost 6 metres (19 feet) long, although a typical size today is under half this length. Alligators are not lizards but descendants of a more advanced group of extinct reptiles called the thecodonts, which were also believed to be the ancestors of the dinosaurs and birds. They differ from more primitive reptiles in having tooth sockets and a hard palate, which enables them to breathe through their noses even when their mouths are full of food, as mammals can.

Along with their relatives the crocodiles, alligators are perhaps the most fearsome of all ambush predators. From the surface, their bodies

LEFT: *The American crocodile can be distinguished from the American alligator by its narrower snout and the fourth tooth from the front on the lower jaw, which protrudes above the "gumline" of the upper jaw. The American crocodile is a very rare animal. There are probably only about 500 left in the brackish bays of the Everglades and the Florida keys.*

OVERLEAF: *Alligators were hunted heavily for their hides until the late 1960s, when federal laws were passed severely curtailing the trade of alligator leather. Between 1870 and 1970, over 10 million alligators were killed in Florida alone. Today, their numbers are recovering and the trading of their hides has resumed on a much reduced scale. In Hong Kong, an alligator-skin briefcase may sell for as much as $7000.*

resemble a floating log. They drift toward their prey with only their nostrils, eyes, and ears clear of the surface. All of these organs are highly effective in air. Alligators are also efficient underwater predators, able to remain submerged for hours without a breath.

Their jaws are enormous and lined with 70 or 80 sharp, conical teeth. With them, they can drag animals as large as a deer or a human underwater, where they attempt to drown them. They share the reptilian inability to chew their food, but they can twist a limb from an animal's body by clamping onto it and rolling in the water. Alligators often cache larger prey and allow their meat to decompose and soften for a week or so to make it easier to swallow.

More commonly, though, their prey are fish and smaller aquatic mammals or birds. In the often murky water where they live, alligators may be able to sense the approach of a fish with small bumps on the scales around their mouths called integumentary sense organs. Some biologists have theorized that these organs may be sensitive to underwater pressure waves.

The most surprising thing about these fearsome reptiles is that the female is a devoted mother. About two months after mating in the spring, she begins building her elaborate nest 3 or 4 metres (10 to 13 feet) from the water. She starts by clearing the ground of vegetation, tearing it out by the roots or cropping it with her jaws. She piles the mulch into a mound using her tail and snout, then forms a crater in the mound by sitting on top of it and rotating on her stomach by pushing with one leg. She then collects mud and aquatic plants and packs them into the crater. Finally, she uses her foot to dig a smaller crater within the mud and vegetation.

Into this cavity she deposits 25 to 60 eggs, usually at night, and then covers them with more mud and vegetation. For about nine weeks, the eggs incubate while the female stands guard nearby, sometimes with her chin resting on the mound. Heat from the composting vegetation in the nest warms the eggs at night.

As with crocodiles and many turtles, alligators have no sex chromosomes. The gender of their offspring is instead determined by the temperature at which the eggs incubate. In the case of alligators, eggs incubated at between 32°C (90°F) and 34°C (93°F) hatch as males. Eggs incubated at between 28°C (82°F) and 30°C (86°F) hatch as females. Between these ranges, the sex ratio is about 50/50.

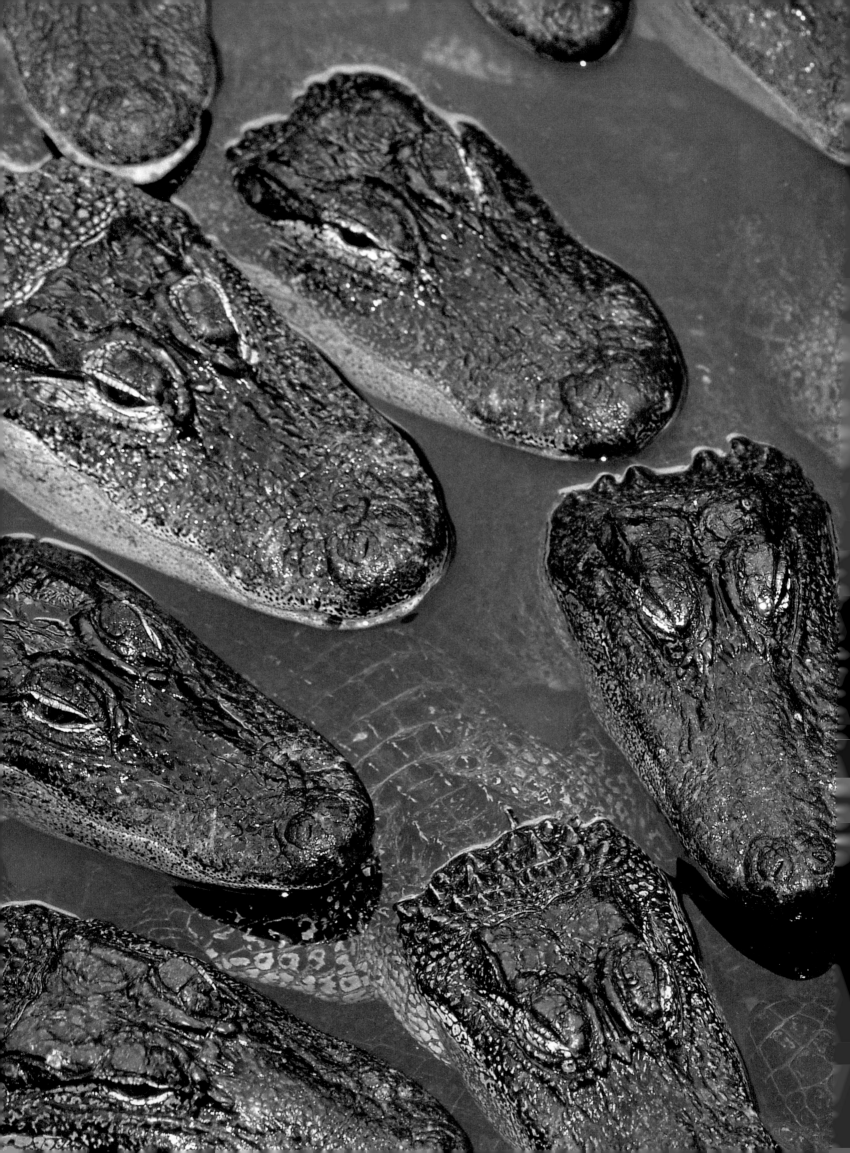

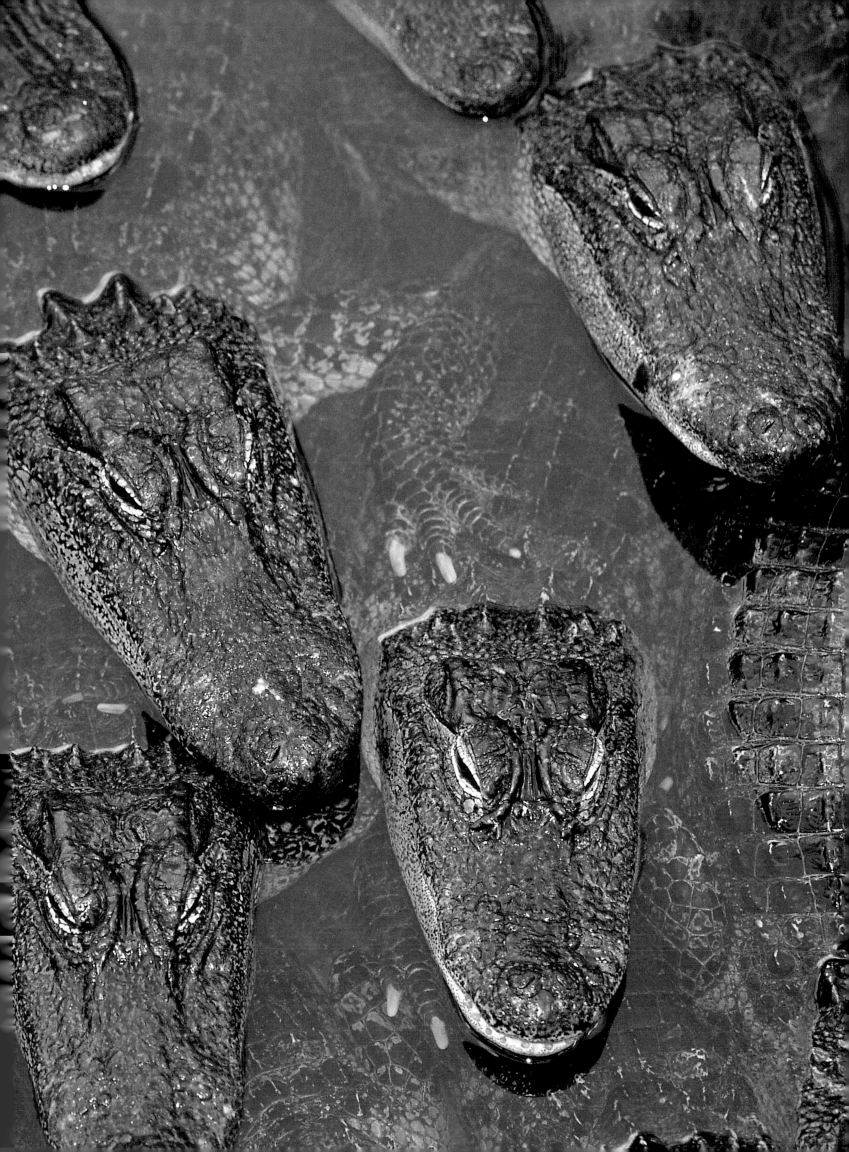

ABOVE: *Some female alligators abandon their nests shortly after laying their eggs, but others are diligent parents. The attentive mother may take her young in her mouth as they emerge from the nest and even rinse them in water to remove bits of shell and amniotic fluid. Occasionally, her back serves as a beachhead during their first few months—the only time in their long lives when they are at all vulnerable to natural predators.*

The gators hatch from mid-August to September. They are about 23 centimetres (9 inches) long, with bright yellow markings on their black bodies. The mother carries them to the water, picking up several at a time with her mouth. They call to her with a strange, gurgling noise that has been used as a sound effect for aliens in dozens of old science fiction movies. If she is any distance from her brood, the female responds by rushing back to them. The young alligators may remain under her protection for up to a year and a half, but hunt for themselves from birth, preying on amphibians, fish, shrimp, and aquatic insects. They may live as long as 40 years.

Alligators are one of the few reptiles to make any sound at all. As adults, they sometimes roar loudly. The sound has been described as thunderous or rumbling, and is sometimes set off by the passage of a truck or heavy equipment. Just prior to emitting their roar, the alligator's torso vibrates. If it's lying in water, these vibrations stipple the water's surface with droplets, as if in a downpour. The display is called "the water dance." Biologists are not sure why they do it, but both males and females bellow.

Alligators are found throughout the southeastern United States, from

North Carolina to Texas. Their range actually extends far enough north that sometimes their ponds may freeze over in winter. At such times, most alligators retreat into dens they dig into dense grass or a mud bank, but a few have been found locked in surface ice—alive—with just their snouts protruding above the water.

Numerous until the arrival of Europeans, alligators were heavily hunted for their leather. Between 1870 and 1970, it's estimated that 10 million were killed. After laws protecting them were passed in the late 1960s, their numbers began to recover, although they are still far from historic levels.

Today, they are threatened by the encroachment of more and more people upon their habitat. Their stronghold is the Everglades, the second largest park in the contiguous United States, which covers most of southern Florida. The Everglades have been described as a river 100 kilometres (60 miles) wide but only 15 centimetres (6 inches) deep. But over the last century, its flow has been strangled by over 2250 kilometres (1400 miles) of dikes, levees, and canals that control floods and supply water for agriculture and the expanding population of southern Florida. Today, only half as much water flows through the river of grass as flowed there a century ago.

One of the most ambitious habitat restoration plans ever devised has recently begun, under the supervision of the National Parks Service and the U.S. Army Corps of Engineers. Over the next two decades, the plan is to restore much of the water that is now diverted out to sea back into the Everglades. Hopes are to return its flow to 70 percent of historic levels.

The project involves the flooding of farm and residential lands, and so is vigorously opposed. It may be enough to save the Everglades or it may not, but at a cost of $8 billion, it is a considerable effort to preserve what is essentially a swamp, an ecosystem dominated by some of the least-loved creatures on earth.

At the very least, it buys the hope that wild places may be valued for their own sake, and that the patience of the reptile is rewarded still.

A Double Life

Most animals have an ideal size, which they attain as adults. Once fully grown, they are large enough to successfully hunt or forage for food, but not so large that they can't find enough to sustain themselves. Until they reach their adult size, many higher animals need their parents' help to feed: Mammals give their offspring milk, and may forage with them for years. Many birds collect food and regurgitate it for their young; by the time they leave the nest, most birds are as large as their parents.

Few amphibians offer their young any care, but they have found a different solution to the problem of surviving early life: Like many of the invertebrates, they live two distinct lives, as larva and adult.

Most frogs, toads, and salamanders hatch from eggs laid in the water, where they then live as tadpoles or larvae. This life may be as short as two weeks for the spadefoot toad tadpole, or as long as two years for a tiger salamander larva. Tadpoles are very much like fish. Their bodies are streamlined—little more than a head with a tail on it—and they can swim rapidly. Their gills take oxygen directly from the water and they have lateral lines—a row of special cells along their sides that are sensitive to

ABOVE: *While other salamanders have a slimy feel, the rough-skinned newt lives up to its name with skin that feels almost prickly to the touch, much like sharkskin. Over most of its range, many potential predators avoid the aquatic, larval stage of this salamander (pictured) because of toxic secretions.*

LEFT: *The bullfrog's large eyes probably have a wider field of view than that of any other vertebrate, except perhaps for some fishes. The eardrums, or tympani, just below and behind the eyes, are used mainly to detect the calls of other bullfrogs. Lower vibrations such as those transmitted by approaching predators may be conducted to the inner ear bones via the frog's forelimbs.*

pressure waves—for detecting the movements of prey and predators. They have tiny beaked mouths, suited for rasping at algae or swallowing free-floating bacteria and other single-celled organisms.

But as they grow, their bodies undergo a metamorphosis in preparation for their second life. They develop limbs for climbing, hopping, and walking on land. Their gills are traded for lungs, so they can breathe out of water. Their mouths grow to almost the width of their bodies, enabling them to eat animals as big as themselves. (In the case of frogs, notorious cannibals, it sometimes *is* themselves.)

Having two lives has been a very effective survival strategy for the amphibians. As aquatic larvae, they avoid many of the predators that would hunt them on land, but as adults they can exploit these same animals as an abundant food source. Their adult mobility also allows them to travel between bodies of water. Many species lay their eggs in a puddle or a hole in the crotch of a tree limb—temporary pools free of fish and other predators. By overwintering in ponds or lakes, some larval salamanders and tadpoles can avoid sub-freezing air temperatures.

This dependence on fresh, liquid water makes most of North America an inhospitable desert for amphibians during much of the year. They are bound to water not just for reproduction, but often for breathing itself. For many species, the uptake of oxygen and release of carbon dioxide occur directly through their skin. Some of the salamanders have no lungs at all. For the dusky salamander of the eastern United States, 85 percent of respiratory exchange occurs directly across the skin, and the rest through the tissues of the throat and mouth. Such an amphibian must keep its skin moist if it is not to suffocate.

Most amphibians also absorb water directly through their skin, and few of them actually drink. Many toads and some frogs living in dry

RIGHT: *The Pacific treefrog shares with all treefrogs somewhat slender legs, and fingers and toes that end in tiny discs. Both are adaptations to a life of climbing rather than leaping. The discs can grip perfectly smooth surfaces, allowing treefrogs to climb windows. Their call, heard day and night, is the peeping so often used to set night scenes on film soundtracks.*

places have an area of thin skin, rich in blood vessels, on the underside of their pelvis that is particularly absorbent. When they come across a puddle or other body of water, they sit in it and soak up the moisture.

Although their delicate skins leave them vulnerable to drying, they are also some amphibians' best defense. The skins of many frogs, toads, and salamanders contain glands that secrete toxins. No amphibian in North America is as deadly as the South American arrow frogs, whose legendary poison, curare, is so powerful that it may kill through skin contact alone, but there are plenty of species that are irritating enough to discourage predators from eating them. Salamanders, frogs, and toads secrete at least 200 known toxins. Some toads secrete a substance that causes snakes to yawn, releasing their prey. Some salamanders secrete a glue that may stick a predator's mouth shut. Others, to judge from the reactions of their would-be predators, just taste bad.

Besides living better through chemistry, amphibians have other defenses. Southern toads swallow air to try to swell themselves beyond the gape of garter snakes. The river frog plays dead when captured. This is thought by some biologists to reduce the chance of it being injured until its toxic skin secretions can take effect on a predator.

ABOVE: *The call of the northern leopard frog is usually described as a low snoring followed by a few seconds of clucking, and has also been compared to the sound produced by rubbing a balloon. People searching for these frogs are often surprised to find that they are calling from underwater, even though they can be heard quite clearly in air.*

LEFT: *The larva of the tiger salamander owes its jesterlike appearance to the external gills branching from its head. These will be lost if the larva metamorphoses into an air-breathing adult. Where terrestrial conditions are harsh and aquatic predators are few, tiger salamanders remain in the larval form their whole lives, a condition known as neoteny.*

ABOVE: *The waterdogs are another group of salamanders that spend their entire lives in the larval state. The dwarf waterdog reaches a maximum length of about 19 centimetres (7 1/2 inches) and lives at the bottom of sluggish streams and ditches, hunting for worms and aquatic insects. It is found from southeast Virginia into Georgia.*

Despite their dependence on water, a number of amphibians have adapted to life in the driest parts of the continent. Some survive in deserts by spending most of the year in a state of torpor, not because it is too cold for them to live, but because it is too dry. This summer hibernative state is called estivation, and is similar to the process in the lungfishes, some of which also live in bodies of water that dry up seasonally. Today's lungfishes may well be descendants of the ancestors of the first amphibians.

During a dry spell, the Mexican treefrog of southern Texas burrows into the soil and then sheds several layers of its skin while secreting mucous between each layer. The mucous and dead skin harden into a form-fitting cocoon. While the living skin of the frog is still permeable to water, the cocoon is not. The frog continues to breathe very slowly through the cocoon's only opening, at its mouth. In this way it can survive months of dry weather, losing very little moisture to the air.

The eastern spadefoot toad also buries itself, with a particular, squirming motion of its body, using the edges of its hind feet almost to screw itself into the ground. Aside from periods encased in a cocoon similar to that of the Mexican treefrog, it spends much of its adult life buried without the special cocoon, relying on the moisture of the soil to keep it from drying out. It may stay this way for weeks at a time.

Spadefoots often emerge from the ground during spring rainstorms, usually at night. Once they have dug themselves out, spadefoots head for a pond, ditch, or other body of water. The males usually arrive first and begin calling to the females, which follow their calls.

The anurans (frogs and toads) are the most vocal of the amphibians. On many summer nights, their calls completely dominate the soundscape. They make their croaks, barks, and peeps by shunting air back and forth between their lungs and extensible throat sacs. As the air moves over the vocal cords, it produces the frog's or toad's characteristic call. The throat sac conserves both water and energy. By keeping the flow of air completely within the body, they lose very little moisture. While it takes more air pressure from the contracting lungs to blow up the throat sac for the first call, on inhalation for the next croak or peep the elastic energy of the throat sac helps to push the air back into the lungs. The throat sac also resonates and amplifies the call. Some frogs call with their throat sacs partially submerged, broadcasting underwater.

BELOW: *Many frogs and toads use abandoned animal burrows to hibernate below the frostline. The wood frog, shown here, has the most northerly range of any North American frog, but survives the winter beneath only a thin covering of leaves or soil, usually blanketed by snow. Before freezing solid, the wood frog floods its body with glucose by breaking down liver glycogen. This prevents the formation of ice crystals that would otherwise damage its cells. Shortly after thawing out, the frogs congregate at ponds and lakes to breed.*

The vocalizations of male frogs serve the same purpose as they do in bugling bull elk or singing robins: They call females to mate and warn other males to keep their distance. Some species have two-note calls, and it's believed by some biologists that one note is the warning and the other the invitation. If a male intrudes on the caller's territory, the caller may switch exclusively to the warning note before attempting to drive off an intruder with physical threats. Some ponds are home to several species of frog, and a distinctive call helps frogs to zero in on mates of their own kind.

These calls can be loud—from 90 to 120 decibels at a distance of 25 centimetres (10 inches). A human being with good hearing could detect a single frog at that volume from almost 2 kilometres (1.2 miles). The frogs, whose ears are naturally tuned to the calls of their own kind, can probably hear them from even farther away. Most frogs and toads have well-developed ears and the drum, or tympanum, is usually visible as a round membrane just behind the eye.

In frog or toad species that have prolonged breeding periods, males often stake out territories. In some species, males alternate their calls with those of their neighbors so that their voices may be distinctly heard, and a female can zero in on an individual frog. Not so with spadefoot toads, which breed frenetically.

Spadefoot males all call loudly, joining their voices to draw as many females as possible. Once the females begin arriving on the breeding grounds, the males scramble for the females and there is little mate selection. The amphibian mating clasp, during which the male clasps the female from behind, is called amplexus. The males of most species grab the female around the forelegs. Spadefoot males clasp their partners around the base of the rear legs.

RIGHT: *The sound of the male bullfrog calling for a mate is one of the earliest signs of spring in North America. A group of scientists in New York has dedicated years to tracking the emergence of the frogs from their winter hibernation, and the beginnings of the distinctive night calls. These researchers believe that the animals are initiating their mating season 10 to 13 days earlier than they did half a century ago — a small sign that global warming may be affecting the intricate rhythms of nature.*

ABOVE: *The spotted salamander spends the winter underground, in either an animal burrow or an existing crevice. In spring, hundreds of them crawl out of the woods to converge on their breeding ponds over one or two nights. Many of the salamanders must cross roads during these mass migrations, and in some areas automobile traffic takes a heavy toll. A spotted salamander's life might otherwise span 20 years.*

As with almost all anurans, fertilization takes place in the water. After the male spadefoot has grasped his partner, the pair dives into the water and the female begins searching for a suitable place to lay her eggs. The male's cloaca is positioned so that as the female extrudes her eggs in a gelatinous string, his sperm enters the water with them. The sperm find their own way to the individual eggs, which usually float in a mass, anchored to a leaf or twig. There may be 2500 eggs in a string, and laying them takes only a minute or two.

Because the spadefoot toad breeds in flooded fields and other temporary bodies of water, the eggs and tadpoles have to develop rapidly. They may hatch in as few as two days after fertilization. For the first few days, the tadpoles remain motionless, clinging to vegetation or other vertical surfaces with a suction cup on their bellies. Over the next three or four days, they swim near the surface, feeding on floating microorganisms. A week or two after hatching, they form dense schools.

Depending on the size of the pond, these schools may be large—one or two square metres (11 to 22 square feet) in area—and contain tens of thousands of individual tadpoles. If the pond is shrinking, crowding may further concentrate them as they compete for plankton. Cannibalism is common among spadefoot tadpoles in a drying pond. Under duress, they may begin to metamorphose into adults in as little as two weeks after hatching. Where water and food are plentiful, it may be two months before they begin sprouting limbs. Almost as soon as they leave the water, the toads disperse into the fields and arroyos from whence their parents came.

The spadefoot toad's life cycle follows a primitive amphibian pattern, but the anurans have developed many variations on this basic plan. The male tailed frog of Washington and southern British Columbia has an extension of its cloaca which it uses as a penis to fertilize the female's eggs internally. This is almost certainly an adaptation to the swift mountain streams where the frogs live. Most of the male's sperm would likely be washed away before it could fertilize the female's eggs if it were simply released into the current. The tadpoles have sucker mouths that allow them to cling to stones in fast-running water. The greenhouse frog, a species introduced to Florida from Central America, lays its eggs in a rotting log or in moist vegetation and the tadpole develops and metamorphoses before it ever hatches. Eventually, it emerges from the egg as a fully formed, though tiny, frog.

As adults, frogs and toads are among the world's best ambush hunters, catching food with a minimum of movement. They can squat on their folded limbs for hours, waiting motionless for insects and other prey to approach. The large eyes, positioned at the sides and tops of their heads, are held high, ideally placed to acquire airborne targets. When it comes, the strike is often a flick of the tongue with almost no other body movement. When frogs and toads are threatened, their folded rear legs uncoil like springs, carrying them many body lengths from their original position in a second—without leaving a particle of scent for a predator to follow.

Considering the variety of habitats in which frogs and toads live, there is remarkably little variation in body shape. They live in swift-flowing streams and buried in desert sands, on the trunks of trees and on the cups of flowers. And yet it would be practically impossible to mistake a frog or a toad living in any of these places for anything else.

Most adult salamanders have a lizardlike body more suited to active hunting. As larvae, they have feathery, external gills. As adults, they crawl through the leaf litter in search of insects, worms, and other invertebrates. Their limbs are weak, and salamanders move by wriggling from side to side, using their legs as levers to pull themselves forward with twists of their trunks.

The mudpuppy is a large salamander up to 50 centimetres (20 inches) long that lives in the Mississippi and Ohio river systems. But the mudpuppy has truncated the normal amphibian life cycle. Like the hellbender and the Pacific giant salamander, the mudpuppy remains in its larval stage for its entire life, retaining its feathery gills, tail fin, and lidless eyes. This condition is called neoteny, and is common to a number of salamanders.

They do, however, mature sexually and so are able to complete their life cycle. Courtship occurs during the fall in deep water at night and so has rarely been observed in the wild. The male walks over and under the female, stroking her body with his own, while she stands rigid on the bottom. The male deposits capsule-shaped globules of sperm called spermatophores about 2.5 centimetres (one inch) long on the river- or streambed near her. She picks them up with her cloaca and the eggs are fertilized internally. It is not until the following spring that she lays as few as 30 or as many as 200 eggs on the underside of a rock ledge or log where she has hollowed out a space. She guards and aerates them for up to two months until they hatch.

The mudpuppy is active mainly at night, and every year someone out fishing reels one in and takes it in to the local university in the belief that he or she has discovered some missing link. In one sense, these would-be Darwins are right: The development of individual amphibians from larva to adult mirrors the evolution of fish to amphibian, and the mudpuppy lives straddling these two animal classes.

The tiger salamander is the largest terrestrial salamander in the world and the most widely distributed in North America, ranging from southeastern British Columbia to Mexico. The larvae can be found in almost any body of water, and the adults in forests, meadows, and even deserts.

LEFT: *The oak toad is North America's smallest toad and, apart from its size, can be identified by the white and orange stripe down the middle of its back. It does almost all of its hunting during the day, and at night it calls from the undergrowth of forests with a peeping, birdlike call.*

ABOVE: *Ensatinas are lungless salamanders represented by several subspecies along the Pacific coast from Vancouver Island to Mexico. The one pictured here is the Oregon ensatina. Its eggs are laid underground or beneath logs and hatch directly into adults, foregoing the aquatic stage entirely.*

RIGHT: *Spadefoot toads, such as these plains spadefoots, lay their eggs in temporary ponds and puddles created by seasonal rains. These have the advantage of being relatively free of predators such as fish or turtles, but often dry up shortly after the rainy season. To cope with this problem, the eggs and tadpoles of spadefoot toads develop rapidly. The eggs hatch within 48 hours and the tadpoles may undergo metamorphosis only two weeks later.*

What's remarkable about the tiger salamander is its variability. In most habitats, it cycles through the usual amphibian life stages over three or four months—egg to larva to adult. But in some places, such as the cold mountains of the west where few of the lakes support fish, and conditions on land are difficult, some of the larvae never mature completely. Like the mudpuppy, they spend their whole lives underwater. There, they are pursued by few predators and remain active throughout the winter in water just above freezing.

Other tiger salamanders remain larvae for a winter or two and then mature. As adults, they hunt through the moist leaf litter for insects, worms, and other invertebrates. Tiger salamanders are seldom seen because they do most of their hunting at night. In the northern reaches of their range, they survive the winter underground, sometimes in burrows abandoned by mammals, but often beneath logs or leaf litter. Following a spring rainstorm, they may migrate from their wintering places by the thousands to the ponds and lakes where they breed. They may live up to 25 years.

Unlike the toads and frogs, the salamanders have experimented with some extreme variations on their basic body plan. The sirens of Florida

and the southeastern United States only ever develop a single pair of legs, the forelegs, and these are tiny in relation to their long eel-like bodies. They hunt for insect larvae and small fish in the mud and among the roots of aquatic plants. The greater siren grows to almost a metre (3 feet) in length. Despite having only the one pair of legs and never losing their gills, sirens occasionally venture onto land.

An equally strange group of salamanders, the amphiumas or mud eels, have four legs but these are so tiny in relation to their rather large bodies that at first glance they appear to have no legs at all. Their gills are enclosed by gill covers, reinforcing the impression of an eel. Like the sirens, they hunt in the muddy sediments of streams and ponds, and rarely leave the water.

The adaptations of amphibians may seem bizarre and extreme to human beings. Imagine a mammal that never grew up, could survive without lungs, or regrow an amputated limb it lost as a juvenile. What we would consider birth defects have evolved into successful ways of life for the amphibians.

In 1976 a species of frog, the gastric brooding frog, was discovered in the Conondale mountain range of Australia. In this species, the female swallows her fertilized eggs and they develop entirely within her stomach. A substance in the jelly surrounding the egg turns off secretion of the mother's digestive acids. During "pregnancy" the mother's lungs collapse to allow the stomach to accommodate the tadpoles, and the frog breathes entirely through its skin. The froglets emerge from the mother's mouth.

Unfortunately, only 20 years after their discovery, biologists can no longer find any of these frogs.

In the last 25 years, for reasons unknown, amphibians worldwide have been in sharp decline. Certainly, populations of mammals, fishes, and

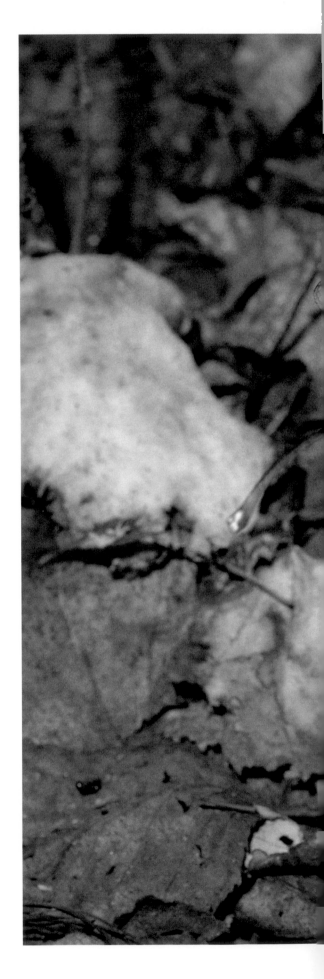

RIGHT: *The American toad is a member of the family of "true toads." All of these toads secrete a milky, white poison from glands on either side of the neck. Most animals trying to ingest a toad will suffer inflammation of the throat and mouth. Absorbed into the bloodstream, the poison causes nausea, irregular heartbeat, and may even kill the would-be predator.*

birds have also suffered from habitat destruction, but the disappearance of the amphibians has been especially alarming. Often, whole populations disappear, while the habitats in which they once lived still seem relatively remote and intact.

Scientists have proposed a number of theories why amphibians, and particularly frogs and toads, may be susceptible to some pervasive change in the environment. Their permeable skins may leave them vulnerable to pollutants. Their eggs may be damaged by the breakdown of the ozone layer and subsequent rise in ultraviolet radiation levels, because they often develop in shallow bodies of water. Or maybe the flood of hormones during the radical reorganization of metamorphosis leaves these animals susceptible to pollutants ingested by the tadpoles.

No one is certain what is happening to the frogs and their relatives; there may be many causes rather than one. But the growing silence of our ponds and marshes turns a key of disquiet in us all.

LEFT: *The broken-striped newt, at home in the marshlands, ponds, and ditches of North and South Carolina, lives its life in three stages. Born as a gilled larva, the newt metamorphoses into a terrestrial eft and leaves the water for one to seven years to feed mainly on insects and worms. After a second metamorphosis, the newt returns to the water to breed.*

A Day at the Beach

The stories we read from the fossil record are like the best fables: impossible to verify, often straining credulity, and yet consistent with all the other tales told by stone. For the marine mammals, the story goes something like this: Some ancestral species of generalized, terrestrial mammal lived by the sea. Life there was hard, usually because of a scarcity of food. But in the sea, food was abundant and so the animals entered the water frequently to feed. Over generations, chance favored some of their offspring with mutations that accumulated to make aquatic life easier: streamlined bodies, more efficient lungs, eyes and nostrils closer to the tops of their heads. Eventually, the animals became more at home in the water than on the land. The marine mammals are like the family who decided to leave their troubles behind them and go to the beach for a day. They found life there so good that they never came back.

The fossil record tends to dwell on its success stories. No doubt there have been mammals, rife with all the genes necessary for an evolutionary sea epic, that waded into the ocean only to be eaten by sharks. No fossil publishers for that short story. But the marine mammals that survive today—the seals, sea lions, walruses, manatees, dolphins, por-

ABOVE: *Like walruses and some whales, the bearded seal sings an unearthly, melodious song while diving beneath the ice of its arctic habitat—probably as part of its courtship behavior. The song can sometimes be heard in the air, and may go on for more than a minute.*

LEFT: *The beluga whale has perhaps the greatest vocal repertoire of all marine mammals. The sounds of this whale have been described variously as the tuning of a string orchestra, oxen lowing, pigs grunting, sighing, a crowd of children at play, trilling, mewling, clucking, and the rattling of sheet metal. The sailors who first heard them through the wooden hulls of their vessels dubbed the belugas "sea canaries."*

poises, and whales—are the culmination of an evolutionary triumph.

For one thing, marine mammals can be huge. Unrestricted by gravity, their bodies have grown far beyond those of terrestrial animals. Size is generally a mark of success for a species: It means that it is good at finding food and has outgrown most of its predators. We think of bears and bison as large animals, but an adult bull bison is only a quarter the weight of an elephant seal.

Their bulk—more specifically, their ability to accumulate blubber—is the key to the marine mammals' success. It allows them to feed in some of the world's coldest and most productive waters, but more importantly, it allows them to *not* feed. Small mammals such as rodents and shrews must have access to an almost continuous supply of food but many of the marine mammals can forego feeding for weeks or months, living off their stores of fat. This turns out to be of great benefit when bearing and nursing their young.

The true or earless seals are not as thoroughly adapted to life in the sea as the cetaceans—the whales and dolphins—in that they must still come ashore to bear their young. The earless seals include the harbor, gray, ribbon, ringed, hooded, northern elephant, and harp seals, which live mainly on the Arctic coasts of Canada and Alaska. On the Pacific coast, harbor and elephant seals range as far south as the Baja peninsula. On the Atlantic coast, harp and gray seals live as far south as Nova Scotia and Maine.

Harp seals breed on ice floes. Because sea ice is frequently shifting and breaking apart, it's a rather dangerous place for a terrestrial animal, and so a haven from predators for seals. There, the only major predators they have to worry about are polar bears and human seal hunters. When one of these predators does find them, seals can slide more quickly over ice

RIGHT: *Researchers have long used the orca's distinctive black and white markings to help identify individual whales. The whales may use them for much the same purpose. At close range in murky water, these contrasting patches may allow orcas to keep track of one another silently during a hunt, when vocalizing would reveal the pod's presence.*

than they can over rock or soil. The water under the ice is often rich with fish, the harp seal's principal food.

But the mothers do not feed at this time. They concentrate on whelping and nursing their pups. The young are born nearly helpless, with none of their mothers' thick layer of blubber. Instead, they have only a thin layer of brown fat, just beneath the skin around the chest and shoulders, which they metabolize rapidly to keep from freezing. Their renowned white fur coats camouflage them from polar bears and insulate them from the frigid arctic air, but the fur is effective only as long as it remains dry. It would be of little help to them in the water, where they will soon spend almost their entire lives.

To help them grow and accumulate their jacket of blubber, the mother pumps them full of her rich milk. Harp seal milk is 20 percent fat (whipping cream is about 30 percent fat) at the pup's birth and increases to almost 40 percent by the time the pup is weaned, just two weeks later. Weaning is without ceremony. The mother, who previously strayed from her pup's side only for brief swims, simply doesn't return one day. The pup, which by now has swelled to 45 kilograms (100 pounds), calls for its mother with a bleating cry, but she never returns. For the next several days, the young seal lies on the ice, pining for its mother and losing almost half its body weight. Eventually, hunger drives the juvenile to the water, where it must learn on its own to hunt. It starts with krill—a free-swimming shrimplike crustacean about the size of a child's finger—and other large plankton, eventually graduating to more difficult prey such as the arctic cod. It is probably the pursuit of schools of these fish that takes harp seals on their annual migrations around the Arctic and north-western Atlantic oceans, journeys of over 9600 kilometres (6000 miles).

The mother swims to nearby ice floes where the males have gathered, waiting for nursing to end. Since the harp seal's gestation period is almost a year, including a four-and-a-half-month period of delayed implantation of the embryo, they mate immediately after weaning their pups. Courtship is a hasty affair in which males chase females over the ice, snapping at their tails, or serenade them from the water, singing and blowing bubbles. They copulate in the water and then resume their solitary lives. And that, essentially, is the social life of a female harp seal: two weeks during which the females bear their pups and attempt to transfer

BELOW: *The northern fur seal is the most pelagic of all seals, spending months at sea and traveling over 10 000 kilometres (6200 miles) in the course of its annual migration. It returns to land only to breed. At sea, it dives for fish and squid, mainly at night when its prey is closer to the surface. By day, fur seals swim lazily at the surface, but must keep moving to generate heat. Their fur, though dense, is not as effective an insulator as blubber.*

as much of their body fat to them as they can, followed by a few days of mating. The males have no contact with their young at all. Although it is not a gregarious existence, over the harp seal's lifespan of about 30 years, it is highly effective at producing more seals. By keeping her pup's help-less, ice-bound youth to a minimum, the mother greatly increases its rate of survival.

It may seem odd that creatures that breathe air should excel at feed-ing in the depths of the ocean, but it has been a source of abundant food for all the seals. A suite of physiological adaptations allows them to pen-etrate the abyss. Just before submerging, a seal exhales most of the air from its lungs to decrease its buoyancy. As its face enters the water, its nostrils close and the heart slows from 100 beats per minute to 5 or 6. Blood vessels to all parts of the body except the heart, lungs, and brain constrict, confining the flow of blood to this essential circuit.

As the seal descends, its highly flexible lungs and ribs are squashed nearly flat by the mounting pressure. The small amount of air left in the lungs is forced up into the bronchi so that little of the nitrogen contained in the air will enter the blood. Were that to happen, it might form bubbles in the seal's blood on its return to the surface, giving it "the bends."

ABOVE: *Steller sea lions continue to suffer a mysterious and distressing decline in their numbers off the Pacific coast. In the 1960s, their range from the Aleutian Islands to the neck of the Baja peninsula was estimated to contain as many as 300 000. Today, there are probably fewer than a third of that number. Declining fish stocks, chemical pollu-tion, and shootings by fishermen, whose nets the sea lion frequently damages, have all been blamed.*

LEFT: *The harp seal pup is born with a white fur coat that will keep it warm for the first few weeks of its life. Before it can enter the water, however, it must accumulate a jacket of blubber, which it does in only two weeks of nursing, swelling to eight times its birth weight.*

All the while, the pressure on the seal's body mounts. For every 10 metres (33 feet) of descent, pressure increases by one atmosphere. At a kilometre (.6 miles) below the surface, where elephant seals have been known to dive, the pressure on every square centimetre of its body exceeds 100 kilograms (1400 pounds per square inch). The overall pressure on the seal's body exceeds that of a junkyard car crusher. The only air spaces inside the seal's body have already collapsed and the rest of the seal's body is mostly water, which is incompressible, and so it cannot be crushed.

When the seal opens its mouth to feed, the throat is blocked by the soft palate and the tongue, and it can even swallow food at this depth. With their blood supply much reduced, the muscles that propel the elephant seal through the water must rely on oxygen contained in the myoglobin of its muscles. This protein, which is similar to blood hemoglobin, is present in much greater concentrations in the muscles of diving mammals than in land mammals.

When it returns to the surface, up to three-quarters of an hour later, the seal's blood flow and heartbeat rise to greater than normal levels until oxygen is restored to its depleted tissues. This may take up to an hour.

The eared seals are the most recently evolved of the seals, and they retain more of the marks of their terrestrial ancestors. These were probably bearlike animals that made their trip to the beach some 30 million years ago. As their name implies, the eared seals still have an external ear flap, although in every other way their bodies are highly streamlined. Unlike adult earless seals with their thick blubber coats, these seals rely on fur for insulation throughout their lives. Northern fur seals have a very dense pelt composed of two kinds of hairs: long, thick guard hairs surrounded by shorter, finer hair fibers. The water penetrates only as far as the guard hairs, which are covered in an oil secreted by glands at the hair

LEFT: *During their long dives, walruses divert the flow of blood away from their skin and blubber to conserve body heat. When they first emerge from the sea their skin is pale, almost white. As the sun warms their bodies, the flow of blood, and their rusty hue, returns to their skin.*

ABOVE: *A sperm whale lifts its massive tail flukes from the water before diving. Sperm whales probably dive deeper than any other air-breathing animal in search of squid, their main prey. Sonar has been used to track them on dives lasting 80 minutes to depths of over 3000 metres (10 000 feet).*

root. Beneath the guard hairs the seal's fine underfur remains dry, providing an insulating jacket. On shallow dives, the fur seal's pelt is an excellent insulator, but on deeper dives, water pressure compresses this jacket of air and fur so that it is not nearly as effective as the blubber of the true seals, walruses, and whales. Despite this disadvantage, northern fur seals spend months at a time without ever leaving the water, relying on almost constant movement of their swimming muscles to maintain their body temperature.

Sea lions are eared seals that have reached a compromise to keep warm. Their coats consist of only one fur layer, but they have far more blubber than a fur seal.

While the true seals can only hump along out of water, the fur seals and sea lions are able to raise themselves on their flippers and waddle on all fours. They can clamber ashore faster and more easily than a true seal. This is particularly helpful when the fur seals are threatened by sharks or killer whales, their only two natural enemies. Once aware of its presence, a seal in water can easily outmaneuver a shark (sharks generally attack marine mammals while their prey is sleeping), but the only escape from the swift and agile killer whale is to scramble onto land.

Although it has no external ears and very little hair, the walrus is more closely related to the eared than the earless seals. Like the eared seals, it has the ability to rotate its hind flippers forward, but its great bulk (an adult bull may weigh up to 1360 kilograms, or 3000 pounds) prevents it from "walking" quite as well. There are two subspecies, Pacific and Atlantic, although not long ago by evolutionary standards there were several Pacific species. Both the Atlantic and Pacific walruses of today are confined to arctic waters, but the Atlantic subspecies once ranged as far south as Sable Island off Nova Scotia. On the Pacific coast, there were walruses in California. The more southerly Atlantic populations were all hunted to extinction by the beginning of the 19th century for the ivory of their tusks, meat, and leather, which could be made into thongs approaching the strength of steel cables. Today, the Pacific population is by far the larger of the two, and their herds are much bigger.

Walruses usually feed in sediments less than 70 metres (230 feet) below the surface. There, they root through the mud, using their whiskers—more properly called vibrissae—to feel for clams, whelks, and sea cucumbers. Once they've found food, walruses may use their tusks to anchor themselves in the mud while squirting away the sediments around the shellfish. They vacuum the clams into their mouths and then, holding the shells in their lips, pull their tongues back within their tube-shaped mouths. The resulting suction pulls the meat of the clam from its shell.

Walruses migrate with the seasonal formation of sea ice, heading north in summer and south in winter, when they mate. The Atlantic males try to defend small herds of females from other males, but the Pacific walruses follow what is known as the lek system of mating. In lek mating, males attempt to impress gatherings of hundreds of females with a courtship display of some kind, and the females gauge their performance to choose a mate.

In both subspecies, the males sing underwater in a voice that has been described as bell-like. While the females congregate on the sea ice, the males swim nearby, wooing them into the water. When a female eventually enters the water to feed, the nearest male usually swims at her side for a time and then tries to clasp her with his foreflippers. She either rejects his embrace or allows copulation, which may last several minutes.

Atlantic walruses are much more aggressive when it comes to driving away rival singing males, which is why they are described as polygynous,

OVERLEAF: *There is no mistaking the sickle-shaped dorsal fin of a killer whale. A male orca's fins and flippers continue to grow throughout its life, and so an old whale can be identified by the proportionately greater size of these appendages. Among the most long-lived of mammals, they may reach 90 years of age.*

or harem-keeping. When one male intrudes on another's territory, the rivals try to intimidate one another with displays of their tusks. The tusks, which grow throughout the walrus's life, are reasonably accurate signifiers of an animal's age and size, so the walrus with the larger tusks is usually declared the victor before a fight can even begin. But males do fight, sometimes by rising out of the water and attempting to drive their tusks into their rival's neck. Walruses have extremely tough skin, especially around the neck and shoulders, but most males bear scars from these fights.

The female walrus gives birth to a single calf in spring. As with seals, the gestation includes a lengthy period of delayed implantation, and lasts more than a year. Walrus childhood is not the rushed affair it is for the harp seal. A pup suckles for up to a year and may remain with its mother for another year after that.

Walruses are gregarious animals outside of the mating season. In summer, the Pacific walrus congregates on sandy or rocky beaches, lying cheek by jowl in herds of thousands. This chumminess probably evolved out of necessity. Beaches both shallow enough for a walrus to climb onto yet still protected from polar bears are at a premium, and therefore crowded.

The Inuit have tales of rogue walruses, animals of exceptional size, yellowed tusks, and foul temper that have taken to eating seals or even beluga whales. Shunned by other walruses, the rogues are said to be more than a match for killer whales, and native hunters use an imitation of the rogue walrus's bellowing call to drive away orcas. Some biologists dismiss the rogue as a myth, and maintain that pockets of walruses have simply taken to eating the meat of marine mammals.

The ability of the seals to feed and accumulate fat under adverse conditions is impressive, but it is dwarfed by the equivalent process in some of the whales. The baleen whales—which include the blue, fin, gray, humpback, sei, and right whales—are the barons of blubber. During summer and fall, most of them feed at the ends of the earth, in the polar seas. These waters are the most productive on earth, nourished by the continuous daylight of the polar summer and minerals dredged from the ocean depths by upwelling currents.

All the baleen whales feed by straining huge quantities of seawater containing plankton—usually krill. The mouth of a blue whale is so large that it can engulf over 65 000 litres (17 000 gallons) of seawater and krill in a single mouthful. The whale strains the water through a comb of baleen hanging from its palate, leaving only the krill and a little seawater in its mouth. Then it swallows. Feeding almost exclusively on these crustaceans—perhaps 40 million of them a day during the feeding season—female blue whales have grown to 190 tonnes, making them the largest animals ever to have lived.

In winter, female baleen whales swim to the warmer and calmer tropics to bear their young. The males follow them and wait for their opportunity to mate sometime after the females have given birth. The females, like harp seal mothers, concentrate on suckling their calves. A

RIGHT: *Humpback whales off the coast of Alaska employ a sophisticated feeding method known as bubble-netting. To concentrate fish, one or more humpbacks dive below a school and swim in a tightening spiral while exhaling. The fish are crowded toward the surface, surrounded by a circle of bubbles. The whales rise beneath the concentrated school with mouths agape.*

ABOVE: *Elephant seals are often plagued by overheating in the summer months and to combat this problem, they shovel sand over their bodies. Once hunted for their blubber, elephant seals came to within 50 animals of extinction in the 1890s. Under a complete hunting ban, their numbers have recovered to almost 100 000.*

mother blue whale will transfer 350 litres (92 gallons) of milk to her calf daily. In the first three weeks of its life, a young blue may gain 113 kilograms (250 pounds)—each day.

In turn, the mother may lose up to a third of her body weight during a winter of nursing. While this is a lot in absolute terms, it is about the equivalent sacrifice of the nursing harp seal mother. During this time, the adult blue whales feed very little, if at all. They fast because it isn't worth their while to feed in the tropical open ocean, which is almost barren of plankton—and because they can.

One of the guiding principles of biology is a law of geometry that dictates the larger an object is, the proportionately smaller its surface area. This has two benefits for the giant whales: For one thing, in comparison to the thrust they are able to develop with their powerful tails, water exerts very little drag on their bodies, and so they swim with little effort. Although they migrate long distances, proportionately it takes them much less energy than it does a dolphin or other smaller, migratory marine mammal. Secondly, a whale, though it may have lost much of its blubber while nursing its young, is in little danger of freezing. It's so big that it is almost impossible for it to lose heat fast enough through its relatively small surface to become chilled even in the coldest water.

The baleen whales have grown so huge as to be almost invulnerable to the main threats to most mammals: starvation or predators. Apart from human beings, the only animals large or powerful enough to threaten the great whales are other whales.

Orcas, or killer whales, are most whales' only natural enemies. There are now thought to be three kinds of orca: resident, transient, and oceanic. The resident whales living in the inshore coast of British Columbia and Washington feed mainly on salmon and other fish. They are somewhat territorial and live in pods of closely related individuals. The transients range up and down the Pacific coast, eating mostly seals, porpoises, and dolphins. They travel in groups of unrelated individuals. The oceanic whales roam the open ocean, and are the ones most likely to attack other whales.

Actually the largest members of the dolphin family, orcas are almost inescapable once they have chosen their prey. They swim as fast as any other whale or dolphin—56 kilometres (35 miles) per hour—and their huge jaws, ringed with banana-sized teeth, are terrible in their efficiency. A large whale, such as a fin or a sperm whale, might break the back of an orca with its powerful tail, but in the open ocean there is no hiding place from the concerted effort of a pod of killer whales, which will take turns tearing at one or two spots on its body until they open a fatal wound. As with wolves, the hunting success of orcas lies in the society of the pack.

So it is with all of the marine dolphins, which coordinate their efforts to hunt and feed upon fish and squid. They are gregarious and highly vocal, signaling to one another with pops, whistles, squeals, and clicks. They use these sounds not only to communicate, but to "see" with sonar, continually mapping their location, that of their pod members, and their prey.

All dolphins and porpoises are members of the toothed-whale family, which also includes the sperm whales, belugas, narwhals, and beaked whales. In contrast to the baleen whales, toothed whales do not alternately gorge and fast. Instead, they dive year round in search of fish and squid, often to great depths. Sperm whales are the deepest divers of all marine mammals, and have been tracked by sonar to depths of 3000 metres (10 000 feet).

Like many of the toothed whales, sperm whales live in matrilineal societies. The females live in pods of about a dozen related individuals—

mothers, sisters, daughters, aunts, and young males. All the adult females share in the care of the young. While a mother dives in search of squid at depths of at least 400 metres (1300 feet), an aunt will "babysit" her calf until she returns. When danger threatens, all the adult whales of the pod surround the young, just as musk oxen do in the high arctic.

In contrast to the baleen whales, toothed whale males are bigger than the females. This size difference reaches its extreme in sperm whales. Bulls may weigh up to 50 000 kilograms (110 000 pounds), while cows reach only a third of their weight.

To reach such sizes, and increase their chances of securing a mate, the males forage in the more productive polar seas in pods of unrelated individuals, half an ocean away from the females for most of the year. Even though they have reached sexual maturity, males younger than 20 years do not even bother to make the annual journey south to mate with the females because their chances of success are poor. Females and the larger bulls will prevent them from mating. Fortunately for these young whales, they have a long reproductive life ahead of them, and many future chances to mate. Sperm whales live up to 80 years.

Of all the marine mammals living off the coast of North America, only one is a herbivore. The West Indian manatee looks a little like a tropical, tuskless walrus with an oversized beaver tail. The resemblance is purely superficial; manatees and walruses evolved from unrelated land animals. Instead of grubbing through arctic mud, the manatees move slowly through the shallow waters off Florida, South Carolina, Texas, and Belize, munching on seaweeds. Seaweed is much lower in calories than shellfish, but through constant grazing these marine mammals can grow just as large as walruses, reaching weights of up to 1360 kilograms (3000 pounds).

LEFT: *Most male mammals hang their testes outside their bodies in a sack to keep them cool, but such a strategy would work against a dolphin's streamlined shape. Instead, nets of blood vessels surrounding the testes carry heat to the surfaces of the dorsal and tail fins, where it is quickly lost to the cooler water.*

Lacking the walrus's coat of blubber, the manatee has none of its tolerance for cold. Many spend their summers in the warm waters of the Gulf of Mexico, but as winter approaches and temperatures in the gulf drop below a balmy 20°C (68°F), they migrate inshore, sometimes swimming into estuaries to reach underwater mineral springs where the water seldom varies from 23°C (74°F).

Hundreds of other manatees congregate around the hot-water outfalls of several electrical generating plants each winter. There, they have become a major tourist attraction. Although divers and boaters are forbidden to approach manatees and discouraged from touching them, these marine mammals are sometimes so friendly that the areas around Homasassa and Crystal springs on Florida's east coast look like underwater petting zoos. The manatees often approach boaters and marinas hoping for a drink of fresh water, and if offered a hose, will take it in their flippers and guide it to their fleshy mouths. In the wild, manatees must periodically enter fresh-water estuaries to drink, but no one knows how often they do this. Their kidneys eliminate some of the salt that they accumulate by swallowing seawater as they feed.

Today, there are at most 2500 manatees off the coast of the United

States, and they may be the world's most closely monitored population of wild animals. Although the manatee has no natural enemies except for sharks, the waters off Florida are filled with pleasure craft that frequently run into them. Ninety percent of manatees bear the scars of these collisions, and it's estimated that 25 percent of them die in boating accidents. Despite complete legal protection, their numbers are barely holding. Manatees have excellent hearing and it is now believed that their eyesight is good, so why they should have such difficulty avoiding boats is unknown. They simply seem to react very slowly to danger of any kind.

Manatees and the related dugongs of the Indian Ocean (another blimp-like marine mammal, but with a mouth like a vacuum-cleaner attachment) are the last surviving families of the order Sirenia. But only 230 years ago, another sirenian lived in the remote Commander Islands, at the far end of the Aleutian arc. George Wilhelm Steller was the only professional naturalist ever to study these giant sea cows. We know of them from his journals of his voyage aboard the *St. Peter,* where he served as the ship's naturalist. The *St. Peter* was wrecked on the return leg of its

BELOW: *Elephant seals dive to great depths in search of their prey— deep-dwelling squid, octopuses, small sharks, and ratfish. A depth gauge attached to an elephant seal accompanied it on a record dive to a depth of 1581 metres (5187 feet). The only air-breathing animals known to dive deeper are whales.*

voyage to North America in 1741, and the crew cast ashore on Bering Island. Steller did not survive the winter following the wreck, but the castaways who were eventually rescued saved his journals.

Steller described the sea cow as resembling an immense seal, perhaps 9 metres (30 feet) long and weighing 10 tonnes, with a head shaped like a buffalo's. It had lips covered with thick bristles and clublike paws with which it pulled itself along the bottom in water so shallow that its back was just awash. It also used its paws to knock kelp and other seaweeds from their holdfasts and these plants were its only known food. Between chewing great mouthfuls of seaweed, it would lift its head from the water to breathe. It was never seen to dive.

Wrote Steller, "Signs of a wonderful intelligence I could not observe, but indeed an uncommon love for one another, which even extended so far that when one of them was hooked, all the others were intent upon saving him."

Fossils of the sea cow dating back 20 000 years have been discovered as far south as California. By the time Steller found them, they were probably already a dying species, hunted to a few thousand animals by native peoples living in its dwindling range. Their history was brought to a rapid close when the shipwrecked men of the Bering expedition discovered that the sea cow's flesh tasted more than a little like beef. From then on, the fur traders hunting the islands for foxes and sea otters exploited them mercilessly. By 1768, the last Steller's sea cow was gone.

Today, all that remains of it are composite skeletons in several museums. The number of sea cows living at the time the Bering expedition found them was probably no more than a few thousand—about the same number of Florida manatees living today. The last of the cold-water sea cows was slaughtered unnecessarily, but that was at least for food. If the manatee meets the same fate, it will be because of an insatiable appetite for pleasure boating.

LEFT: *The West Indian manatee is North America's only herbivorous marine mammal. These large, sluggish animals graze on aquatic vegetation and live only in warm water. In winter, they often congregate near the coolant outfalls of power plants. Despite having a face that only a mother could love, these endangered mammals have become the darlings of Florida's tourist trade.*

Index

Bold entries refer to photographs.

Photo Credits

Victoria Hurst 1, 6, 11, 19, 34, 35, 38, 44, 130, 196, 201, 205

Thomas Kitchin 3, 13, 16, 18, 20, 21, 22, 43, 48, 49, 50, 53, 54, 58, 59, 63, 70, 73, 86, 89, 94, 117, 119, 129, 131, 146, 154, 166, 173, 175, 178, 190, 206, 250, 264, 282, 286, 289

Jerry Kobalenko / First Light 5, 187

Mark Degner 8, 90, 106, 108, 162

David Nunuk / First Light 9, 257

Tim Christie 12, 15, 25, 32, 40, 47, 56, 72, 74, 81, 97, 101, 105, 120, 168, 171, 180, 188, 208

Victoria Hurst / First Light 14, 31, 37, 164

Thomas Kitchin / First Light 26, 61, 64, 66, 114, 116, 182, 224, 274, 277

Lynn M. Stone 29, 55, 83, 121, 125, 126, 140, 147, 149, 161, 172, 181, 185, 198, 202, 203, 214, 236, 241, 242, 246, 259, 263, 266, 271, 272, 281, 292

Wayne Lynch 41, 65, 69, 76, 78, 92, 96, 102, 123, 151, 153, 158, 176, 177, 189, 192, 227, 228, 229, 232, 235, 238, 243, 245, 249, 261, 269, 275, 280, 288

Brian Milne / First Light 45, 107, 111

Robert Lankinen / First Light 60, 82, 124, 209

Jason Puddifoot / First Light 77

Tom Ellison / First Light 84

Robin Brandt 87, 186, 279, 296

Craig Brandt 98, 103, 112

Jerry Kobalenko 110

Brent Matsuda 132, 139

Milton Rand / First Light 133

© 1999 B. Matsuda / AAA Image Makers 135, 142

Rod Planck / First Light 136

Barry Mansell 138

Joe & Carol McDonald / First Light 144

David Jones 150

Thomas Kitchin / Victoria Hurst 156, 225, 254, 255, 258, 268

Leslie Degner 193

Ron Sanford / First Light 195

Joe McDonald / First Light 210, 222, 231

M. B. Fenton 211, 213, 220, 221

David M. Dennis / First Light 217, 260

Jerry Lee Gingerich, DVM 218

Gary McGuffin / First Light 252

K. Aitken / First Light 278, 291

Flip Nicklin / First Light 284, 294

Chris Cheadle / First Light 297

Tom & Therisa Stack / First Light 298